AS

AQA BIOLOGY
Specification B

BIOLOGY

Revision and Summary Book

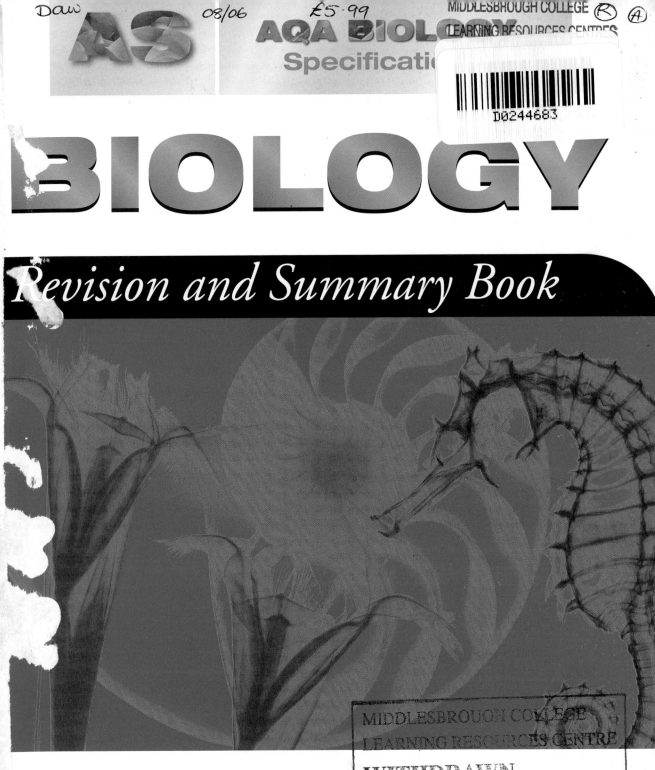

Margaret Baker
Martin Rowland

Hodder Murray
A MEMBER OF THE HODDER HEADLINE GROUP

Photo credits and Acknowledgements

Thanks are due to the following copyright holders for permission to reproduce photographs:

Fig. 1.2a NIBSC/SPL; Fig. 1.2b Science Photo Library; Fig. 1.3 Andrew Syred; Fig. 1.6 Dr Jeremy Burgess/SPL; Fig. 1.7 DR Gopal MURTI/SPL; Fig. 1.8 J.L. Carson, Custom Medical Stock Photo/SPL; Fig. 9.1b Andrew Syred/SPL; Fig. 9.2 Manfred Kage/SPL; Fig. 12.1a Andrew Syred/SPL.

Every effort has been made to contact copyright holders but if any have been inadvertently overlooked the Publishers will be pleased to make the necessary arrangements at the earliest opportunity.

AQA(NEAB)/AQA examination questions are reproduced by permission of the Assessment and Qualifications Alliance.

Orders: please contact Bookpoint Ltd, 130 Milton Park, Abingdon, Oxon OX14 4SB. Telephone: (44) 01235 827720. Fax: (44) 01235 400454. Lines are open from 9.00–6.00, Monday to Saturday, with a 24 hour message answering service. Email address: orders@bookpoint.co.uk

You can also order through our website www.hoddereducation.co.uk

British Library Cataloguing in Publication Data
A catalogue record for this title is available from the British Library

ISBN 0 340 813571

First published 2004
Impression number 10 9 8 7 6 5 4 3
Year 2010 2009 2008 2007 2006 2005

Illustrations by Art Construction
Typeset in 10pt Goudy by Tech-Set Ltd.
Printed in Spain for Hodder Murray, an imprint of Hodder Education, a member of the Hodder Headline Group, 338 Euston Road, London NW1 3BH

Hodder Headline's policy is to use papers that are natural, renewable and recyclable products and made from wood grown in sustainable forests. The logging and manufacturing processes are expected to conform to the environmental regulations of the country of origin.

Contents

Introduction

About this book

This is not a traditional textbook and will not substitute for one. Instead, it should be used as a revision aid, to help consolidate your recall and understanding and to develop examination skills. The chapters have been written to a common style and they share the following features.

CHAPTER OPENING

Each chapter opens with a summary of what you should know or be able to do. You can use this as a quick reminder of each topic, to reinforce what you know and can do or to prompt you to do further revision.

THE TEXT

Most biology teachers extend the subject beyond the specification content to help you put information in a broader context or to interest you and motivate your learning. We have not done this: this book contains only what is in Specification A of the Assessment and Qualification Alliance (AQA) and nothing else. If you look at questions in past examination papers that test recall, you will find the answers are in this book. If examination questions go beyond the content of this book, the examiner will have given you the additional information you need within the question. Do not be put off by long examination questions: they usually indicate that skills other than recall are being tested. Questions testing recall are usually very short.

We have kept the amount of continuous prose to a minimum. Instead, we have presented information in ways that we know students find easy and quick to use. These include lists of bullet points, tables and annotated diagrams. Not only will this help you to structure your revision of facts, processes and principles but, by using the tables and diagrams, it will help you to develop interpretive and analytical skills that are tested in examinations. Wherever possible, we have used diagrams that you could reproduce yourself in an examination.

EXAMINER'S TIPS

Between us, we have considerable experience of setting and marking AS examination papers. Using this experience, we have included tips in each chapter that will help you to improve your examination performance. In these tips, we point out errors committed by past candidates or inform you of good practice.

IN-TEXT QUESTIONS

Examinations test skills, so you need to develop these as you learn and revise. Each chapter has in-text questions that encourage you to become actively involved in your learning. Reading the text without interacting with what you have just read is not an effective way to learn or revise.

Answers have been provided for all the in-text questions at the back of each chapter. Try to answer the questions yourself and use our answers only to check your own. Remember, you must focus on developing the skills that will be tested in your examinations.

EXAMINATION QUESTIONS

Each chapter contains questions from past AS examinations that are reproduced or adapted by permission of the Assessment and Qualifications Alliance (AQA).

At least one question in each chapter contains a student's answer that has been marked by us. We have also given a commentary to explain why the candidate's answer gained, or failed to gain, marks. The more you learn to think like an examiner, the better your grade will be.

Good luck – we look forward to marking your scripts.

1 Cells and cell structure

After revising this topic, you should be able to:

▶ describe the major differences between prokaryotic cells and eukaryotic cells and between animal cells and plant cells

▶ show understanding of the principles and limitations of transmission and scanning electron microscopes, and distinguish between the terms resolution and magnification

▶ identify and explain the functions of the major cell organelles in eukaryotic cells and, where appropriate, show understanding of how the functions of these organelles are interdependent

▶ identify and explain the functions of the following features of a bacterial cell: capsule, cell wall, genetic material

▶ show understanding of the preparation of temporary mounts and of simple staining techniques

▶ estimate the actual size of cells and of cell structures from photographs and drawings

▶ explain how cell components can be separated by cell fractionation followed by ultracentrifugation, and interpret the results of such processes.

Prokaryotic and eukaryotic cells

All cells have the following features:
- an outer barrier, the plasma membrane, which controls the entry of substances into and out of the cell
- cytoplasm, which does most of the 'work' of the cell
- genetic material, made of DNA, which controls the activities of the cytoplasm.

Cells are classified according to the nature of their genetic material.
- Prokaryotic cells have their genetic material free in their cytoplasm. Only bacteria have prokaryotic cells.
- Eukaryotic cells have their genetic material arranged in chromosomes within a nucleus. Animals and plants have eukaryotic cells.

You need to be familiar with prokaryotic and eukaryotic cells and of the main similarities and differences between them. These are summarised in Figure 1.1 and in Table 1.1.

EXAMINER'S TIP
Do not write about 'bacterial chromosomes' since bacteria do not have them. Instead, write about the genetic material of bacterial cells. You would gain credit for stating that 'the genetic material of bacteria is not arranged in chromosomes' if asked to describe a difference between prokaryotic and eukaryotic cells.

| Feature | Prokaryotic cell | Eukaryotic cell | |
		Animal cell	Plant cell
Size	0.5 to 5.0 μm	up to 400 μm	up to 400 μm
Capsule	Often present – may offer protection against attack by other cells	Absent	Absent
Cell wall	Present (made of peptidoglycans)	Absent	Present (made of cellulose)
Plasma membrane	Present	Present	Present
Genetic material (made of strands of DNA)	Circular, not covered by proteins and found free in the cytoplasm	Linear, covered by proteins and found in a nucleus	Linear, covered by proteins and found in a nucleus
Membrane-bound organelles in cytoplasm	Absent	Present	Present
Large, fluid-filled vacuole in cytoplasm	Absent	Absent	Present

TABLE 1.1 The main features of bacterial cells, animal cells and plant cells. Bacteria have prokaryotic cells whereas animals and plants have eukaryotic cells.

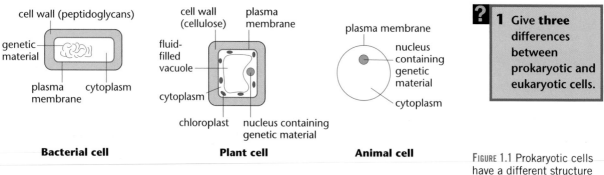

Bacterial cell   **Plant cell**   **Animal cell**

Not drawn to scale

1 Give three differences between prokaryotic and eukaryotic cells.

FIGURE 1.1 Prokaryotic cells have a different structure from eukaryotic cells.

EXAMINER'S TIP

You might be asked in a unit test to identify differences between prokaryotic and eukaryotic cells. Make sure you give different *types* of examples. For example, you would gain 1 mark for stating that prokaryotic cells lacked the membrane-bound organelles that are present in eukaryotic cells or that they lacked a named example of a membrane-bound organelle found in eukaryotic cells. You would *not* gain three marks for giving three examples of membrane-bound organelles that are present in eukaryotic cells but absent in prokaryotic cells.

Seeing cells

To see cells clearly, you need to magnify them using a microscope. Since cells are usually colourless, their structure is best seen if they are treated with chemicals that react with their internal structures.

TYPES OF MICROSCOPE

Table 1.2 summarises the differences between optical microscopes (light microscopes) and electron microscopes. Two terms are used in Table 1.2 which you must understand:

◆ Magnification. This is the extent to which the image is larger than the object being viewed. It is calculated as apparent size divided by actual size. For example, if the image you see is 20 times larger than the actual object, the magnification is 20×.

◆ Resolution. This is the ability of a microscope to distinguish between structures that are very close together. Beams of electrons have smaller wavelengths than do beams of light. Beams of electrons can pass between objects that are about 1 nm apart; their resolving power is, therefore, about 1 nm. Beams of light can only pass between objects that are about 20 μm apart; their resolving power is, therefore, about 20 μm.

> **EXAMINER'S TIP**
>
> Do not confuse resolution with magnification. Make sure you can explain the two terms: resolution is harder to explain simply.

Feature	Optical microscope	Electron microscope
Radiation used	Light	Beams of electrons
Method of focus	Glass lenses	Electromagnets
Maximum magnification	Up to 1500×	Up to 500 000×
Maximum resolution	20 μm	1 nm
Advantages and limitations of use	◆ Relatively easy to use ◆ Preparation of specimen is simple to perform and unlikely to distort actual cell structure ◆ Can be used to view living cells ◆ Have poor power of resolution, so show little detail of cell ultrastructure	◆ Difficult to use ◆ Preparation of specimen is complex and likely to distort actual cell structure ◆ Cannot be used to view living cells since they must contain a vacuum (to avoid deflection of electrons by air particles) ◆ Have high power of resolution, so show much greater detail of cell ultrastructure

TABLE 1.2 A comparison of optical and electron microscopes

? 2 Define the term resolution used in microscopy.

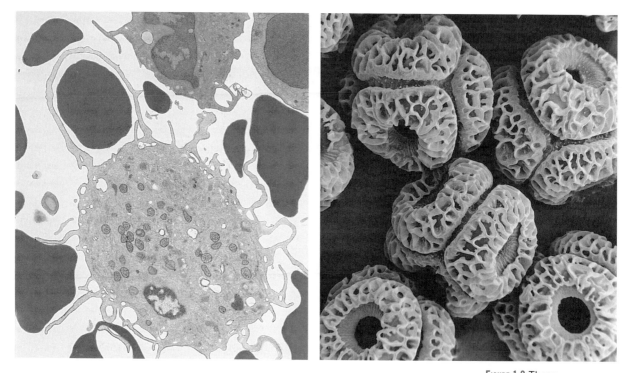

Two different types of electron microscope were used to produce the electron micrographs shown in Figure 1.2.

◆ Transmission electron microscopes pass beams of electrons through the specimen.
  – Since electrons are easily deflected, these specimens must be extremely thin.
  – Following treatment by heavy metal solutions, some parts of the specimen prevent electrons from passing through. These appear dark in electron micrographs.
◆ Scanning electron microscopes bounce beams of electrons from the surface of the specimen.
  – The specimens need not be thin.
  – The magnification and resolution of scanning electron microscopes is usually much less than those of transmission electron microscopes.

FIGURE 1.2 These photographs were taken using a transmission electron microscope (left) and a scanning electron microscope (right).

PREPARATION OF TEMPORARY MOUNTS OF SPECIMENS

Before they can be viewed under an optical microscope, specimens are usually:
◆ cut into thin slices
◆ mounted – slices are fixed on to a glass slide
◆ treated with chemicals that react differently with different parts of the cell (e.g. coloured stains).

? 3 Suggest **two** reasons why structures seen in electron micrographs might not accurately represent those in living cells.

FIGURE 1.3 Staining these onion epidermal cells with iodine solution has made their structure clearer. (×200)

Figure 1.3 shows a temporary mount of onion epidermal cells stained with iodine solution. You should be able to identify cell walls, nuclei and cytoplasm in these cells. Table 1.3 summarises the stages that must be followed in preparing a temporary mount.

Procedure	Explanation
Place a drop of water on a glass slide.	A thin film of tissue will float flat on the surface of the water.
Peel the thin film of cells from the inner surface of the fleshy leaf of an onion bulb. Cut a 5 × 5 mm square of this film and float it on the drop of water on your slide.	The thin film is a single layer of epidermal cells. Done properly, you will have a single layer of cells (with no folds) floating on the water. float film of cells on drop of water slide
Use a mounted needle to lower a glass coverslip slowly over the layer of cells. coverslip mounted needle	The coverslip slows dehydration. The needle is used to avoid trapping air bubbles in the water. Bubbles will look like black circles under the microscope.
Place a drop of iodine solution at one edge of the coverslip and draw the iodine solution under the coverslip using a piece of filter paper. filter paper iodine solution iodine solution drawn under coverslip	The iodine solution will stain some parts of the plant cells, making them easier to see.
Blot away any surplus fluid from the slide. View the slide under a microscope, first with a low-power objective lens and then with a high-power objective lens.	Removing surplus water protects the microscope. The cells are easier to find using the low-power objective lens.

TABLE 1.3 The stages you will follow in producing a temporary mount of stained plant tissue

ESTIMATING THE SIZE OF SPECIMENS UNDER THE MICROSCOPE

When you view a specimen using a microscope, it has been magnified. If you have used a 30× objective lens and a 10× eyepiece lens, the image is 300× bigger than the specimen (30 × 10). However, this does not tell you the size of the specimen.

You do not need to be able to measure the actual size of a specimen, only to estimate it. The simplest way to estimate the size of an object under the microscope is to:

- ◆ measure the diameter of the field of view with the low-power objective lens by observing a plastic ruler or a stage micrometer under the microscope
- ◆ focus on the object in question and estimate its length in relation to the diameter of the field of view you have measured
- ◆ repeat if using the high-power objective lens.

? 4 Explain why describing magnification as 300× is not a satisfactory method of estimating the size of a specimen.

? 5 Using a low-power objective lens, you have measured the diameter of the field of view as 2 mm. You estimate that the length of a structure occupies one-quarter of the diameter of the field of view. What is the estimated size of this structure?

CALCULATING THE SIZE OF SPECIMENS IN A UNIT TEST

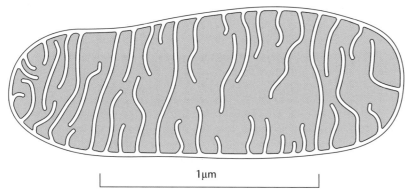
1µm

FIGURE 1.4 A drawing that could be used in an exam to test your ability to measure and calculate the size of structures seen using a microscope.

Figure 1.4 shows a drawing that you might be given in an exam. The drawing includes a scale bar, indicating the actual size of the specimen. To calculate the size of the specimen you must:

- ◆ use a ruler to measure the length of the scale bar (61 mm)
- ◆ use a ruler to measure the length of the cell structure under consideration (109 mm)
- ◆ divide the measured length of the cell structure by the measured length of the scale bar (109/61 = 1.8)
- ◆ multiply this value by the units given on the scale bar (1.8 × 1 µm = 1.8 µm). You must ensure you are able to repeat this calculation with other similar questions from past unit tests.

EXAMINER'S TIP

Questions testing your ability to calculate the size of a specimen are common in BYA1 exams. It is worth practising them.
- ◆ Examiners expect you to measure very accurately, so take your time in making, and checking, your measurement.
- ◆ Remember to take a ruler, graduated in millimetres, and a calculator into your unit test. Without them, you cannot make any measurements or perform accurate calculations.

DRAWING SPECIMENS SEEN UNDER A MICROSCOPE

In your coursework, you are required to draw a specimen (Skill E). Often, candidates are given a specimen which must be viewed using an optical microscope. Such a drawing could be:

◆ a tissue plan, e.g. of the tissues in a leaf, seen using the low-power objective lens of an optical microscope

◆ a drawing of individual cells seen using the high-power objective lens of an optical microscope.

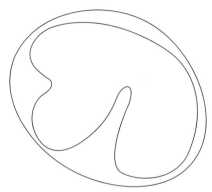

FIGURE 1.5 A drawing that would gain all 3 marks for Skill E in your coursework

Figure 1.5 shows a drawing that would gain all 3 marks available for Skill E. Note that the drawing should:

◆ accurately represent the shape, location and proportion of the actual specimen material – your teacher will verify this by comparing your drawing with the specimen that you observed

◆ be a simple line drawing, free of shading or other artistic embellishment

◆ be drawn with a freshly sharpened pencil.

EXAMINER'S TIP

◆ When drawing a tissue plan, e.g. in a section through a blood vessel, do not attempt to draw individual cells. Instead, draw the regions where different cell types are found.

◆ If you are drawing cells, e.g. in the epithelium of alveoli, draw only a small number of them and make sure you draw them accurately.

Cell ultrastructure

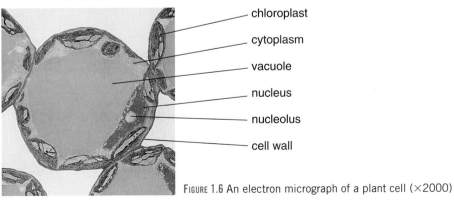

- chloroplast
- cytoplasm
- vacuole
- nucleus
- nucleolus
- cell wall

FIGURE 1.6 An electron micrograph of a plant cell (×2000)

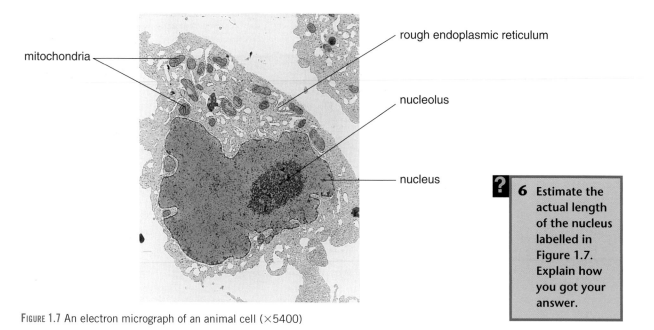

mitochondria

rough endoplasmic reticulum

nucleolus

nucleus

FIGURE 1.7 An electron micrograph of an animal cell (×5400)

? **6** Estimate the actual length of the nucleus labelled in Figure 1.7. Explain how you got your answer.

Figures 1.6 and 1.7 are electron micrographs of a plant cell and an animal cell, respectively. You must be able to identify those structures that are labelled. You must also understand their functions. Table 1.4 summarises what you need to know about these structures.

Cell component	Description	Function
Cell wall	◆ Plant cells only ◆ Layers of cellulose fibres found around plant cells	◆ Support ◆ Prevents cells bursting in dilute solutions
Plasma membrane	Double layer of phospholipids (see Chapter 3); appears as two dark lines	◆ Controls movement of substances into and out of cell ◆ Allows cell identification
Microvilli microvillus	Folds in the plasma membrane	Increase surface area to volume ratio of cells involved in exchange of materials

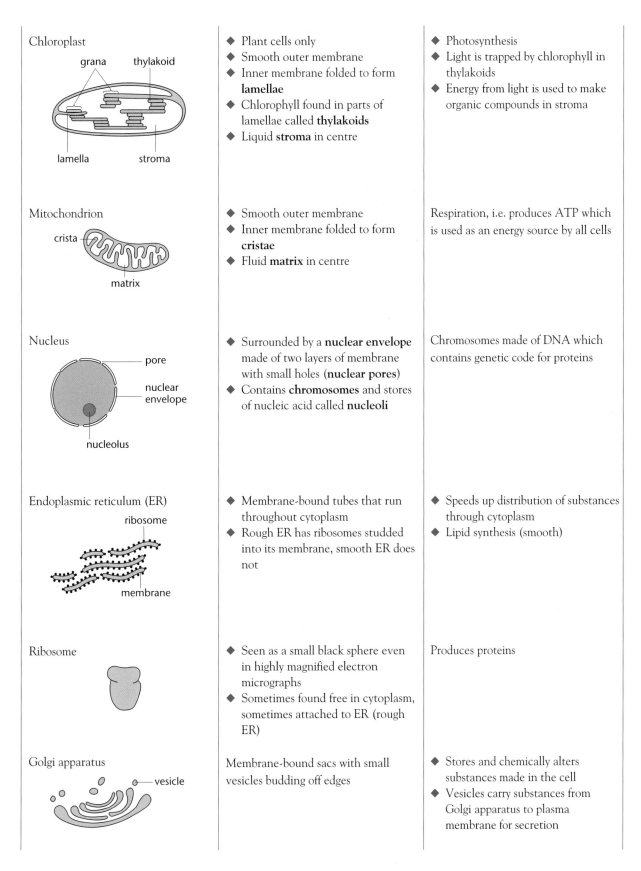

Chloroplast

grana thylakoid

lamella stroma

- ◆ Plant cells only
- ◆ Smooth outer membrane
- ◆ Inner membrane folded to form **lamellae**
- ◆ Chlorophyll found in parts of lamellae called **thylakoids**
- ◆ Liquid **stroma** in centre

- ◆ Photosynthesis
- ◆ Light is trapped by chlorophyll in thylakoids
- ◆ Energy from light is used to make organic compounds in stroma

Mitochondrion

crista

matrix

- ◆ Smooth outer membrane
- ◆ Inner membrane folded to form **cristae**
- ◆ Fluid **matrix** in centre

Respiration, i.e. produces ATP which is used as an energy source by all cells

Nucleus

pore

nuclear envelope

nucleolus

- ◆ Surrounded by a **nuclear envelope** made of two layers of membrane with small holes (**nuclear pores**)
- ◆ Contains **chromosomes** and stores of nucleic acid called **nucleoli**

Chromosomes made of DNA which contains genetic code for proteins

Endoplasmic reticulum (ER)

ribosome

membrane

- ◆ Membrane-bound tubes that run throughout cytoplasm
- ◆ Rough ER has ribosomes studded into its membrane, smooth ER does not

- ◆ Speeds up distribution of substances through cytoplasm
- ◆ Lipid synthesis (smooth)

Ribosome

- ◆ Seen as a small black sphere even in highly magnified electron micrographs
- ◆ Sometimes found free in cytoplasm, sometimes attached to ER (rough ER)

Produces proteins

Golgi apparatus

vesicle

Membrane-bound sacs with small vesicles budding off edges

- ◆ Stores and chemically alters substances made in the cell
- ◆ Vesicles carry substances from Golgi apparatus to plasma membrane for secretion

Vesicle 	Membrane-bound sphere containing fluid	See Golgi apparatus
Lysosome 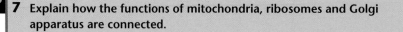	A vesicle that contains powerful proteases called **lysozymes**	◆ Digest protein, e.g. unwanted organelles ◆ In some white blood cells (Chapter 11) they are used to digest bacteria

TABLE 1.4 The major organelles found in eukaryotic cells

> **? 7** **Explain how the functions of mitochondria, ribosomes and Golgi apparatus are connected.**

EXAMINER'S TIP

> You will not be given credit for an answer which states 'mitochondria produce energy'. Instead, write 'mitochondria produce ATP, which is an energy source for cells'.

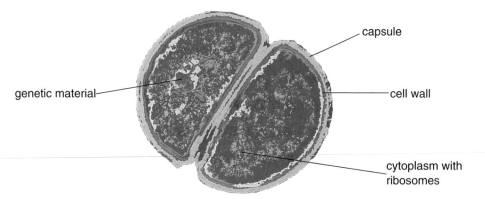

FIGURE 1.8 An electron micrograph of a bacterial cell

> **? 8** **Explain the appearance of the lighter regions in Figure 1.8.**

Figure 1.8 is an electron micrograph of a bacterial cell. The structures that are labelled are those that you must be able to identify. Bacteria have prokaryotic cells. Remember you must be able to distinguish between prokaryotic cells and eukaryotic cells (see Figures 1.6 and 1.7). The main differences can be found in Table 1.1.

SEPARATING CELL COMPONENTS FOR FURTHER STUDY

This technique is used to collect large numbers of a single type of cell organelle. It involves the stages shown in Table 1.5.

Stage	Process	Explanation
Cell fractionation	Cells suspended in ice-cold, isotonic, buffer solution	◆ The low temperature slows enzyme action ◆ The solution must be isotonic so that osmosis does not distort or burst the organelles ◆ A buffer is used so that the activity of the organelles is not affected by changes in pH
	Cells broken up using a pestle and mortar or an electric blender	Frees the organelles from the cells
	Filtration	Separates debris from the organelles
Ultracentrifugation	Filtered solution spun in a centrifuge	Depending on the speed and time of centrifugation, pellets containing different organelles are formed at the bottom of the centrifuge tube (see Table 1.6)

TABLE 1.5 Separating cell components

Speed of ultracentrifuge/ g	Time of centrifugation/ minutes	Organelles in pellet
500 to 1000	10	Nuclei and chloroplasts
10 000 to 20 000	20	Mitochondria and lysosomes Chloroplasts (in plant cells)
100 000	60	Ribosomes

TABLE 1.6 Separation of organelles following cell fractionation (g is the acceleration due to gravity and has a value of $9.8\,m\,s^{-2}$)

9 During fractionation, why are cells suspended in a solution that is: a) ice-cold and b) of the same water potential as the cells?

10 In an investigation, a biologist homogenised liver tissue. After centrifuging the filtered homogenate at 1000 g for 10 minutes, she threw away the pellet and centrifuged the remaining fluid supernatant at 15 000 g for 20 minutes.
a) Suggest which cell organelles she might have been interested in isolating.
b) Explain why she first centrifuged the filtered homogenate at 1000 g for 10 minutes and threw away the pellet.

EXAMINER'S TIP

Remember that organelles are separated by size during ultra-centrifugation, the largest first and the smallest last. The sequence from largest to smallest is nuclei, chloroplasts, mitochondria, lysosomes and, finally, ribosomes.

WORKED EXAM QUESTION

1 a) The drawing was made from an electron microscope. It shows some microvilli on an epithelial cell from the small intestine.

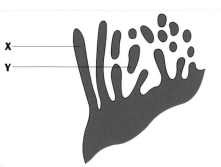

   (i) A transmission electron microscope uses beams of electrons. Explain how a beam of electrons allows the microvilli to be seen in detail. *(2 marks)*

   The one electron microscope magnifies things better than a light microscope so their detail can be seen clearly.

   > The candidate has given an explanation in terms of magnification. The real advantage of an electron microscope is its resolving power. The two marks would be gained for: electrons have a short wavelength; this gives high resolution.

   (ii) Explain why the microvilli labelled **X** and **Y** differ in appearance.
   (1 mark)

   Y was broken during cutting but X was not.

   > This is not an adequate answer. The microvilli would not all be in a flat plane: X was but Y was not. The correct answer is: they were cut through a different plane.

   b) Different cells contain different numbers of mitochondria. Suggest the advantage of large numbers of mitochondria in

   (i) a cell from a plant root which absorbs mineral ions from the soil;

   Mitochondria produce energy which is needed for active transport of ions in the root.

   > Mitochondria do not 'make' or 'produce' energy – they make ATP from which energy can be released. The statement about energy being needed for the active uptake of ions is correct, though, and gains a mark.

   (ii) a muscle cell. *(3 marks)*

   Energy is needed for the contraction of muscles.

   > This answer is correct and gains a mark. The candidate has not gained the third mark for this question, which would be awarded for stating that mitochondria make ATP.

(AQA 2003)

EXAMINATION QUESTION

1   The diagram shows a section through part of a cell as it would appear when seen with an electron microscope.

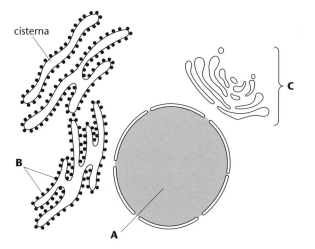

a) This cell produces and secretes a protein. Describe the part played by organelles **A**, **B** and **C** in producing and secreting this protein. *(3 marks)*

b) The table shows information about the different parts of this cell.

Part of cell	Percentage of cell volume	Number in cell
Cytoplasm surrounding cell organelles	54	1
Mitochondria	22	about 1700
Nucleus	6	1
Lysosomes	1	about 300
Cisternae of rough endoplasmic reticulum	9	1

(i)   Which organelle is larger, a mitochondrion or a lysosome? Use calculations based on figures from the table to support your answer. *(2 marks)*

(ii) In the drawing there appear to be a number of separate cisternae in the rough endoplasmic reticulum. The table gives the approximate number of cisternae as one. Suggest an explanation for the apparent difference. *(2 marks)*

(iii) This cell produces a large amount of protein. Explain how the number of mitochondria in the cell may be linked to this. *(3 marks)*

(AQA 2002)

2   Getting into and out of cells

After revising this topic, you should be able to:

▶ show an understanding of the arrangement of phospholipids, proteins and carbohydrates in the fluid-mosaic model of membrane structure

▶ show an understanding of the ways in which a plasma membrane acts as a barrier to the movement of molecules and ions

▶ show an understanding of the process of diffusion across membranes and use Fick's law to explain the factors that affect the rate of diffusion across membranes

▶ explain facilitated diffusion across a membrane in terms of carrier proteins in the membrane

▶ use the term water potential to explain osmosis and to interpret the movement of water from one cell to its environment or to another cell

▶ explain active transport across membranes in terms of trans-membrane proteins and ATP hydrolysis

▶ show an understanding of endocytosis and exocytosis.

Living cells must take in metabolites, such as oxygen and nutrients, and excrete metabolic wastes, such as carbon dioxide. Cells also produce materials for secretion, such as enzymes and hormones. In order to enter or leave the cell, these substances have to pass across the plasma (cell surface) membrane. Before you can understand these processes, you need to understand the structure of the plasma membrane that surrounds every cell.

Structure of plasma membranes: the fluid-mosaic model

? **1** Define the term metabolite.

Every cell is enclosed by a plasma membrane which acts as a barrier to the free passage of most molecules. The components of every plasma membrane are shown in Table 2.1.

Component of plasma membrane	Description	Effect on movement of substances across membrane
Phospholipid bilayer	A double layer of phospholipid molecules. Each molecule has a hydrophilic 'head' and a hydrophobic 'tail'. These molecules are only stable when mixed with water if they form a bilayer.	Phospholipid prevents the movement of any water-soluble ion or molecule but allows lipid-soluble substances to diffuse through the membrane.

Proteins embedded in the phospholipid bilayer	Extrinsic proteins are found only on one side of the phospholipid bilayer.	Act as receptor sites, combining with specific substances that have a complementary shape to their own. Do not affect the movement of substances.
	Intrinsic proteins span the whole membrane.	Help water-soluble molecules and ions to cross the membrane.
Carbohydrates	Attach to some proteins on the outside of the membrane to form glycoproteins.	Involved in receptor sites and do not affect the movement of substances.

TABLE 2.1 The components of plasma membranes

Figure 2.1 is a static representation of a plasma membrane. In reality, the molecules in a plasma membrane move about.

◆ They move laterally about the membrane.
◆ They leave the membrane to join vesicles in the cell's cytoplasm.
◆ They join the membrane from vesicles in the cell's cytoplasm.

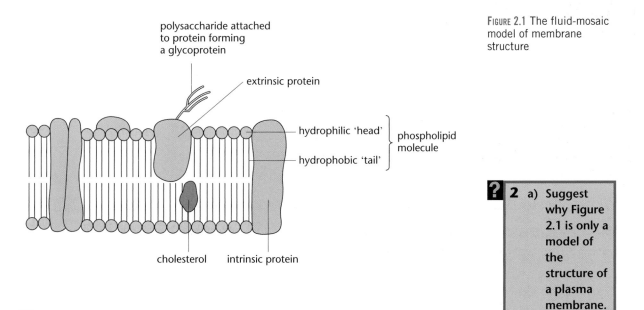

FIGURE 2.1 The fluid-mosaic model of membrane structure

2 a) Suggest why Figure 2.1 is only a model of the structure of a plasma membrane.
b) Explain why the plasma membrane in this model is described as a fluid mosaic.

Movement across membranes

A plasma membrane is partially permeable. This means that it will only allow the movement of some substances across itself, into or out of the cell.

◆ Large molecules, such as proteins, cannot cross membranes because their molecules are larger than the pores in the membrane.

◆ The hydrophobic tails of the phospholipids act as a barrier to water and to water-soluble substances. Only lipid-soluble substances can dissolve in the phospholipid and diffuse through the bilayer.

◆ Intrinsic proteins allow water-soluble molecules and ions to cross membranes. Different types of intrinsic protein act as:
  – **carriers** for water-soluble molecules (such as glucose)
  – **channels** for ions (such as sodium ions and chloride ions)
  – **pumps** to move water-soluble molecules and ions using energy from ATP hydrolysis.

> **? 3** Give **four** functions of the proteins in a plasma membrane.

DIFFUSION

Molecules and ions move randomly about. They move more in gases than in liquids and, in both, move more at higher temperatures than at lower temperatures. During this random movement, they tend to spread themselves about until they are uniformly distributed in the available space. This is called **diffusion**, which is:

◆ the net movement of molecules or ions from a region of their high concentration to a region of their low concentration until the concentrations are equal (see Figure 2.2)

◆ a passive process, i.e. it does not require energy from the hydrolysis of ATP.

> **EXAMINER'S TIP**
>
> The pathway from a region of high concentration to a region of low concentration is called a **concentration gradient**. If you use this term in an exam, make sure you refer to diffusion *down* a concentration gradient. You will not be given credit for referring to diffusion along a concentration gradient since you will not have made clear the direction of movement.

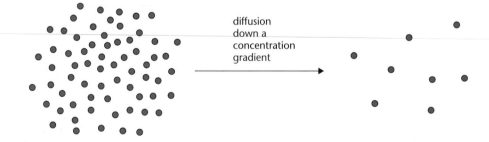

diffusion down a concentration gradient

Figure 2.2 Diffusion is the net movement of ions or molecules from a region where they are concentrated to an area where they are less concentrated, i.e. down a concentration gradient. This is a passive process that continues until the concentrations are equal.

Rate of diffusion

The rate of diffusion across an exchange surface is increased if:

◆ the surface area of the exchange surface is large in relation to its volume

◆ there is a large difference in concentration across the exchange surface (you could state the same idea as 'the concentration gradient is large')

◆ the distance across the exchange surface is small.

The relationship between these factors is summarised in Fick's law, which is given below.

Diffusion rate is proportional to $\dfrac{\text{Surface area} \times \text{Difference in concentration}}{\text{Thickness of exchange surface}}$

Fick's law does not cover all the factors that affect the rate of diffusion across biological exchange surfaces. In addition, the rate of diffusion will be high if:

◆ the temperature is raised, giving the molecules or ions more kinetic energy
◆ the diffusing molecules or ions are small so that they can pass easily through pores in the membrane
◆ where relevant, there is an increase in the number of channels that allow diffusion of the molecules or ions in question in the plasma membrane of cells lining the exchange surface.

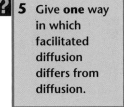

EXAMINER'S TIP

In a unit test, you might be asked to show a qualitative understanding of this law, i.e. you will *not* be asked to carry out calculations involving Fick's law.

> **4** Insulin affects the plasma membrane of liver cells. It causes protein channels that allow the passage of glucose to join the surface membrane from the cytoplasm.
> a) Suggest why insulin affects cells in the liver but not cells in the heart.
> b) Describe and explain the effect of insulin on the rate of glucose diffusion into a liver cell.

Facilitated diffusion

Facilitated diffusion occurs through protein molecules that span the plasma membrane. In other respects, it is like diffusion in that it is a passive process and will only take place down a diffusion gradient across the membrane. Substances that are not lipid soluble generally diffuse across membranes by facilitated diffusion. They include:

◆ relatively large molecules (such as glucose or amino acids)
◆ charged ions.

Figure 2.3 shows two types of protein that are involved in facilitated diffusion.

◆ Carrier proteins bind to a specific type of molecule (such as glucose) on one side of the membrane. They then change their shape and release the molecule on the other side of the membrane.
◆ Ion channels are formed by proteins with a central pore and can be open or closed to regulate the flow of charged ions (such as Na^+ and Cl^-).

> **5** Give **one** way in which facilitated diffusion differs from diffusion.

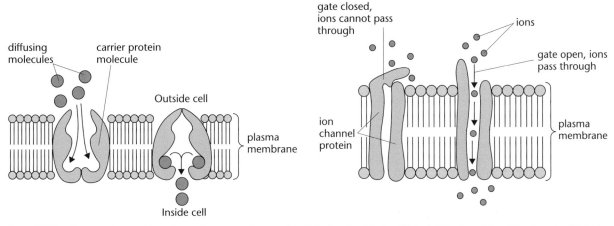

FIGURE 2.3 Carrier proteins and ion channels are two types of protein involved in facilitated diffusion. Like diffusion, facilitated diffusion is a passive process.

OSMOSIS

Osmosis is the diffusion of water molecules across a partially permeable membrane (see Figure 2.4). To explain this adequately, you must understand two concepts: partially permeable membranes and water potential.

◆ A partially permeable membrane is one that allows water molecules to pass through it but will not allow larger molecules, such as proteins, to pass through it. Water molecules are not lipid-soluble, so they cannot dissolve in, and cross, phospholipid bilayers. They must cross plasma membranes through specific protein channels. It is these channels that cause partial permeability.

◆ Water potential is a measure of the pressure that water molecules exert on a membrane. It is represented by the Greek letter psi (ψ).

   – During their random motion, water molecules hit plasma membranes. This causes a pressure on the membrane.

   – The greater the number of water molecules in a solution, the greater the number of times water molecules hit a plasma membrane, causing a greater pressure and so increasing the water potential.

   – Pure water has more water molecules per unit volume than any solution. Therefore, its water potential has the highest value possible: it is given a value of zero.

   – Since the water potential of all solutions will be less than that of water, they have a negative value. Like all pressures, water potential is measured in MPa.

Remember:

◆ water has a high water potential – its value is zero

◆ a solution has a lower water potential than water – its value is negative.

> **? 6** Solution A has a water potential of −1.5 MPa and Solution B has a water potential of −0.5 MPa.
>   a) Which solution has a water potential nearer that of pure water?
>   b) Which solution has the higher water potential?

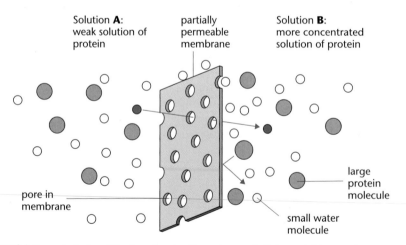

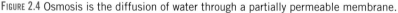

FIGURE 2.4 Osmosis is the diffusion of water through a partially permeable membrane.

> **EXAMINER'S TIP**
>
> During osmosis, water molecules diffuse down a water potential gradient. You could correctly write this in one of two ways, i.e.:
> ◆ there is a net movement of water molecules from a solution of high water potential to a solution of lower water potential
> ◆ there is a net movement of water molecules from a solution with a less negative water potential to a solution with a more negative water potential.

Figure 2.5 summarises the process of osmosis.

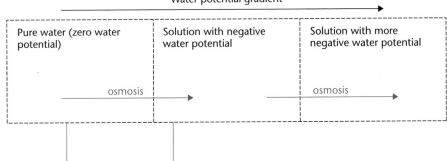

FIGURE 2.5 A summary of osmosis

7 If the two solutions in Question 6 above were separated by a partially permeable membrane, in which direction would osmosis take place. Explain your answer.

EXAMINER'S TIP

Candidates often show confusion in questions testing their understanding of osmosis.
Remember that:
- pure water has the highest water potential, $\psi = 0$
- solutions always have a lower water potential than pure water, $\psi = -$value
- water will always pass by osmosis from pure water to any solution.

Water passes by osmosis from zero ψ to negative ψ and from less negative ψ to more negative ψ.

You should remember that all plasma membranes are partially permeable. Table 2.2 shows that osmosis can have a significant effect on cells.

Cell type	Cell placed in solution with a more negative water potential than its own cytoplasm (a hypertonic solution)	Cell placed in solution with a less negative water potential than its own cytoplasm (a hypotonic solution)
Animal cell	Water leaves cell by osmosis. Cell contents decrease in volume. Cell shrinks and appears crinkled.	Water enters cell by osmosis. Cell contents increase in volume and exert pressure on plasma membrane. Cell bursts because plasma membrane cannot withstand this pressure.
Plant cell	Water leaves cell by osmosis. Cell contents decrease in volume and shrink away from cell wall. A cell in this condition is said to be plasmolysed.	Water enters cell by osmosis. Cell contents increase in volume and exert pressure on plasma membrane and on cell wall. Cell does not burst because wall exerts a restraining inward pressure. A cell in this condition is said to be turgid and further entry of water by osmosis stops.

TABLE 2.2 The effect of osmosis on animal cells and on plant cells

? **8 a)** A plant cell has a water potential (Ψ) of -4 MPa. If this cell were placed in a solution with a water potential of -1 MPa, would water move into or out of the cell? Explain your answer.

b) Bacteria have cell walls. Some types of antibiotic stop bacterial cells from producing new cell walls when they divide.
  (i) Suggest why these antibiotics are effective in killing bacteria.
  (ii) Suggest why these antibiotics do not harm human cells.

ACTIVE TRANSPORT

Active transport:

♦ is the movement of molecules or ions from where they are less concentrated to where they are more concentrated, i.e. up a concentration gradient

♦ relies on specific carrier proteins within plasma membranes to transport the molecules or ions concerned

♦ uses energy released by the hydrolysis of ATP, summarised in the equation below.

$$\text{ATP} \xrightarrow{\text{ATPase}} \text{ADP} + \text{P}_i + \text{energy}$$

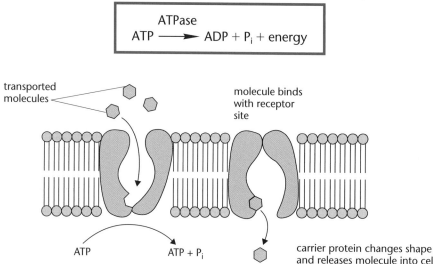

transported molecules

molecule binds with receptor site

ATP   ATP + P$_i$ carrier protein changes shape and releases molecule into cell

FIGURE 2.6 Active transport involves trans-membrane proteins and energy that is released from the breakdown of ATP.

Figure 2.6 summarises the process of active transport.

♦ The specific carrier protein is activated by ATP and binds to the type of molecule or ion that it transports.

♦ The carrier protein then changes shape, releasing this molecule or ion on the other side of the membrane.

The reabsorption of glucose from the proximal convoluted tubule in the kidney is one example of active transport.

EXAMINER'S TIP

Candidates often confuse active transport with facilitated diffusion as both processes rely on specific protein molecules in the plasma membrane. Remember that, unlike active transport, facilitated diffusion does *not* require energy from the hydrolysis of ATP and does *not* transport substances up a concentration gradient.

? **9** Give **two** ways in which active transport differs from facilitated diffusion.

ENDOCYTOSIS AND EXOCYTOSIS

All the processes described already transport only one or a small number of molecules or ions at a time. Endocytosis and exocytosis move substances into, and out of, the cell in bulk. Figure 2.7 shows how both processes use vesicles.

◆ Endocytosis is the bulk transport of large particles (phagocytosis) or of fluids (pinocytosis) into the cytoplasm of a cell from its surroundings. During endocytosis, part of the cell surface membrane sinks into the cell. This part then 'buds off' from the surface membrane and seals back on to itself. This produces a vesicle containing substances from outside the cell.

◆ Exocytosis is the bulk transport of large particles or fluids out of the cytoplasm of a cell into its surroundings. Vesicles are formed from the Golgi apparatus in the cytoplasm of the cell. These fuse with the surface membrane, releasing their contents outside the cell.

Endocytosis

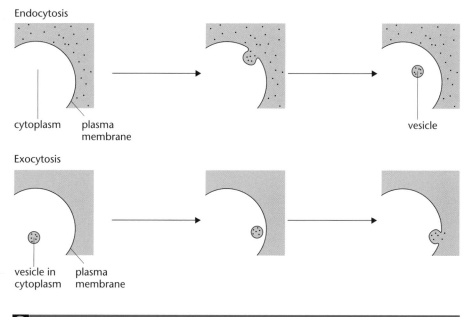

cytoplasm   plasma
             membrane

vesicle

Exocytosis

vesicle in   plasma
cytoplasm   membrane

FIGURE 2.7 Endocytosis and exocytosis involve the bulk transport of substances between cells and their environment.

10 Use Figure 2.7 and your knowledge of cell organelles to answer the following questions.
   a) Had the substances being released from the cell during exocytosis actually crossed the plasma membrane? Explain your answer.
   b) Had the substances being taken into the cell during endocytosis actually crossed the plasma membrane? Explain your answer.

WORKED EXAM QUESTIONS

1   A plasma membrane surrounds an animal cell. Cell membranes are also found in the cytoplasm. The table shows the distribution of membranes around and in the cytoplasm of two different types of animal cell.

Type of membrane	Percentage of total cell membrane	
	Cell A	Cell B
Plasma membrane surrounding cell	2	5
Rough endoplasmic reticulum	35	60
Golgi apparatus	7	10
Outer mitochondrial membrane	7	4
Inner mitochondrial membrane	32	17

a)   Explain why the figures for cell A do not add up to 100%.   *(1 mark)*

The table does not include cell A's nuclear membrane.

> The candidate has used an incorrect term – nuclear membrane instead of nuclear envelope – but has got the correct idea, i.e. that other cell organelles are not included in the table.

b)   (i)   Cell **A** takes up large amounts of substances by active transport. Explain the evidence from the table which supports this statement.   *(2 marks)*

Active transport uses energy which is made by mitochondria. Cell A has more mitochondrial membrane and so can make more energy.

> The candidate gains one mark for identifying that cell A has more mitochondrial membrane than cell B. The candidate has incorrectly written that mitochondria make energy. He would have gained a second mark for stating that mitochondria produce ATP or that active transport moves substances against/up a concentration gradient.

(ii)   Cell **B** synthesises large amounts of enzymes. Explain the evidence in the table which supports this statement.   *(3 marks)*

Enzymes are proteins which are made on the rough endoplasmic reticulum. Cell B has more rough ER than cell A.

> The candidate gains all three marks for: cell B has more rough ER; protein synthesis occurs on rough ER; enzymes are proteins.

c)   Describe the structure of a phospholipid molecule and explain how phospholipids are arranged in a plasma membrane.   *(6 marks)*

A phospholipid molecule is made of a molecule of glycerol joined to two fatty acid molecules and a phosphate group. The fatty acids form a hydrophobic tail and the phosphate forms a hydrophilic head.
The phospholipids are arranged in two layers in the membrane, like a sandwich. Because the tails will not mix with water they are on the inside of the sandwich and the heads are on the outside, like the bread.

The candidate has chosen to use a sandwich to illustrate his answer, but the meaning is clear. This brief answer gains the full six marks, showing that giving a concise answer works! Points are gained for: phospholipid has glycerol; two fatty acids; phosphate; arranged as bilayer in membrane; heads on outside and tails on inside (of bilayer); heads hydrophilic and tails hydrophobic.

2   The graph below shows how the rate of entry of molecules into a cell varies with the external concentration of the molecules. Curve **A** is a case of simple diffusion and curve **B** refers to facilitated diffusion.

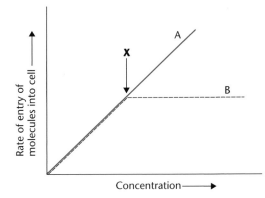

a)   Explain why the shapes of the curves are similar up to point **X**.   (*2 marks*)

Simple diffusion and facilitated diffusion are quicker if there is a large concentration gradient. So up to point X the rate of entry will increase as the concentration increases. This is true for both simple diffusion and facilitated diffusion.

A good answer for 2 marks.

b)   Explain why the shapes of the curves are different after point **X**.   (*2 marks*)

In simple diffusion, the rate at which molecules enter the cell will continue to increase as the external concentration increases. Facilitated diffusion requires energy and after point X the amount of energy is decreasing, so the rate of facilitated diffusion is decreasing.

The first point is correct for one mark. However, in the second part of the answer the student has confused facilitated diffusion with active transport and does not gain the second mark. The correct answer would have been 'in facilitated diffusion the rate slows down when the protein carrier molecules become saturated with the transported molecules and therefore cannot work any faster'.

(AQA 2003)

EXAMINATION QUESTION

1 a) Describe a chemical test you could carry out to show that a piece of coconut contains lipids.   (3 marks)

   b) The diagram shows the structure of a phospholipid molecule.

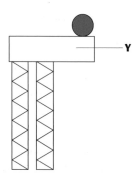

   (i)   Name the part of the molecule labelled **Y**.   (1 mark)

   (ii)   Describe how a phospholipid molecule differs in structure from a triglyceride molecule.   (1 mark)

   (iii) Chitin is a nitrogen-containing polysaccharide. Name **one** chemical element present in a phospholipid which would not be present in chitin.   (1 mark)

   c) An artificial membrane was made. It consisted only of a bilayer of phospholipid molecules. In an investigation, the permeability of this artificial membrane was compared with the permeability of a plasma membrane from a cell. Explain why:

   (i)   both membranes allowed lipid soluble molecules to pass through;   (1 mark)

   (ii)   only the plasma membrane allowed glucose to pass through.   (2 marks)

(AQA 2001)

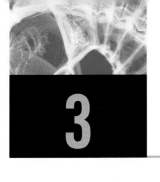

3 Biological molecules

After revising this topic, you should be able to:

▶ state that biological molecules frequently consist of monomers and that these monomers are often combined into polymers

▶ explain that polymers are made by condensation reactions and broken down by hydrolysis reactions, and use this information when interpreting the formation and breakage of glycosidic bonds and peptide bonds

▶ recall, with understanding, the structure of:
  - carbohydrates: α and β glucose, starch, glycogen, cellulose
  - proteins: general structure of an amino acid; primary, secondary, tertiary and quaternary structure of proteins
  - lipids: saturated and unsaturated triglycerides, phospholipids

▶ show understanding of how the structures of cellulose, glycogen and starch molecules and of the tertiary structure of a globular protein are related to their functions

▶ describe, and interpret the results of, tests for the presence of reducing sugars, non-reducing sugars, starch, proteins and lipids

▶ describe how chromatography can be used to separate and identify the components of a mixture; calculate and use Rf values.

Monomers and polymers

Biological substances contain few chemical elements, the most common being carbon, hydrogen, oxygen and nitrogen. Atoms of these elements are chemically bonded together to form molecules. The simplest molecules are called monomers, which can often join together to form chains, called polymers (see Figure 3.1).

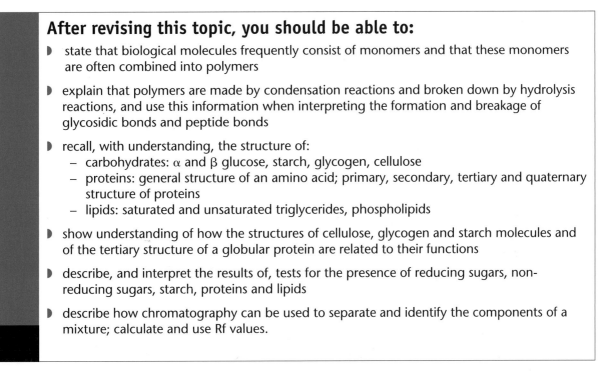

monomer

two monomers
bonded together

chemical
bond

polymer

FIGURE 3.1 Individual biological molecules often join together to form larger molecules called polymers.

Figure 3.2 shows that:

◆ two monomers join together in a condensation reaction – a reaction that eliminates a molecule of water
◆ a monomer is removed from a polymer in a hydrolysis reaction – a reaction that uses a molecule of water.

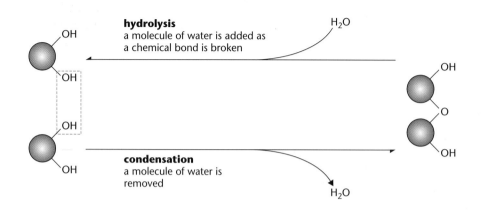

EXAMINER'S TIP

Candidates often forget to show the water molecule in their equations of condensation and hydrolysis reactions. Make sure you show it.

FIGURE 3.2 Monomers join together by condensation (a molecule of water is lost from the joining monomers). Monomers can be reformed by a hydrolysis reaction (a molecule of water is used to separate the monomers).

Carbohydrates

Carbohydrates contain only carbon, hydrogen and oxygen. They have the general formula $C_xH_{2y}O_y$. Table 3.1 shows a number of carbohydrate monomers (called monosaccharides) and the polymers made from these monomers. You must be able to recall these compounds and their functions.

Type of carbohydrate	Examples from specification	Component monomers	Biological importance
Monomers (called monosaccharides, but this is *not* a specification term) ◆ triose ($C_3H_6O_3$) ◆ pentose ($C_5H_{10}O_5$) ◆ hexose ($C_6H_{12}O_6$)	Triose Ribose Deoxyribose Glucose		Respiration and photosynthesis Sugar in RNA Sugar in DNA Used in cell respiration as a source of energy
Two linked monomers (called disaccharides, but this is *not* a specification term) Have the formula $C_{12}H_{22}O_{11}$	Maltose	α-glucose + α-glucose	Sugar obtained from the hydrolysis of starch
Chains of linked monomers (called polysaccharides, but this is *not* a specification term)	Cellulose Glycogen Starch	β-glucose α-glucose α-glucose	Found in plant cell walls Storage carbohydrate in human muscles and liver Storage carbohydrate in plant seeds and other storage organs

TABLE 3.1 Types of carbohydrates and their importance

CARBOHYDRATE MONOMERS: E.G. GLUCOSE

Glucose is an important 6-carbon sugar. It is also the monomer from which cellulose, glycogen and starch are made.

◆ The formula for glucose (and for all other 6-carbon sugars) is $C_6H_{12}O_6$. Compounds with the same molecular formula (e.g. $C_6H_{12}O_6$) are called isomers.
◆ The chemical differences between the different 6-carbon sugars result from the way that their carbon, hydrogen and oxygen atoms are arranged.
◆ Glucose itself has two isomers. These are called α-glucose and β-glucose and their structural formulae are shown in Figure 3.3.

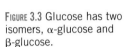

1 If glucose has the formula $C_6H_{12}O_6$, why is the formula for maltose $C_{12}H_{22}O_{11}$ rather than $C_{12}H_{24}O_{12}$?

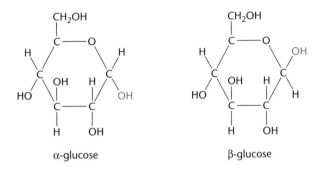

FIGURE 3.3 Glucose has two isomers, α-glucose and β-glucose.

EXAMINER'S TIP

It is important that you know the structural difference between α-glucose and β-glucose. In β-glucose, the hydrogen atom on carbon 1 is in the lower position (it might help to remember 'hydrogen is below' (β low)).

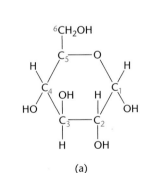

(a)

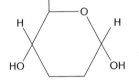

(b) simplified version of diagram (a)

FIGURE 3.4 Diagram (a) shows the structural formula of a glucose molecule. The small coloured numbers enable us to refer to particular carbon atoms in the molecule. Diagram (b) shows a simplified version of this molecule and is all you need to learn for a unit test.

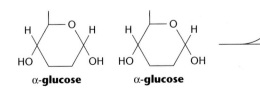

FIGURE 3.5 Two molecules of α-glucose can join together in a condensation reaction to form a disaccharide, called maltose. The bond formed in the reaction is a glycosidic bond.

Figure 3.4 shows a molecule of α-glucose in which the carbon atoms have been numbered. Carbon atoms numbered 1, 4 and 6 are important when monosaccharides join together.

Figure 3.5 shows how two molecules of α-glucose join together.
- ◆ The reaction is a condensation reaction.
- ◆ The new bond formed between the monosaccharides is called a glycosidic bond.
- ◆ Since the glycosidic bond is formed between carbon atom 1 of one glucose molecule and carbon atom 4 of the other glucose molecule, it is called a 1,4-glycosidic bond.

> **? 2** Why is the reaction between two monosaccharides called a condensation reaction?

CARBOHYDRATE POLYMERS

Glucose can form large polymers, called polysaccharides. You only need to know about three – cellulose, glycogen and starch.
- ◆ Starch and glycogen act as stores of glucose.
  - – Starch is found in plant cells.
  - – Glycogen is found in animal cells.
- ◆ Cellulose strengthens the cell walls of plants.

Starch contains two types of chains of α-glucose:
- ◆ amylose, in which all the glucose units are joined by 1,4-glycosidic bonds. This makes their chains straight (see Figure 3.6).
- ◆ amylopectin, in which most of the glucose units are joined by 1,4-glycosidic bonds but some are joined by 1,6-glycosidic bonds. The latter bonds introduce branches into amylopectin chains (see Figure 3.6).

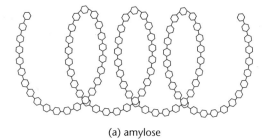

(a) amylose

(b) amylopectin

FIGURE 3.6 (a) Amylose contains straight chains of α-glucose. These chains often coil into a helix. (b) Amylopectin contains straight chains of glucose with some branched chains.

Starch is an ideal store of glucose for the following reasons:
- ◆ It is insoluble.
  - – It will stay in one place, such as starch granules in the stroma of chloroplasts and in the cytoplasm of plant cells.
  - – It will not affect the water potential of the cell storing it. This means that, unlike sugars, stored starch will not cause water to enter the storage cell by osmosis.
- ◆ It contains a large number of glucose monomers, which will provide a large amount of respiratory substrate when hydrolysed.
- ◆ Its molecules have a helical shape, making it a compact store – large amounts of starch can be stored in a small space.
- ◆ It is easily hydrolysed at its ends to release maltose.
  - – Glucose can be made by hydrolysing maltose and can then be used in respiration.

Glycogen molecules are similar to amylopectin but their chains of α-glucose are even more branched. As a result, glycogen has the same advantages as starch as a store of glucose, plus it is more rapidly broken down than starch since it has more 'ends' that can be hydrolysed.

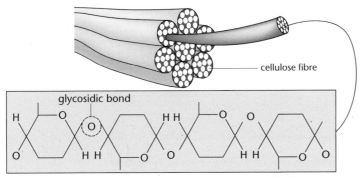

cellulose fibre

glycosidic bond

Long chain of 1,4 linked β-glucose residues forms a microfibril

FIGURE 3.7 Cellulose is a polymer of β-glucose molecules, linked by 1,4-glycosidic bonds. Bundles of cellulose molecules are important in making the cell walls of plants strong.

Cellulose is made from long, straight chains of β-glucose, linked together by 1,4-glycosidic bonds. It is found in the cell walls of plant cells. Figure 3.7 shows how, in plant cell walls:

◆ individual cellulose molecules are joined along their length by hydrogen bonds to form microfibrils
◆ parallel microfibrils are held together in fibres. The fibres in one layer of the wall run in the same direction but at a different direction to those in another layer of the wall.

A plant cell wall contains many layers of cellulose fibres, with the fibres in each layer running in different directions. This arrangement of cellulose fibres gives strength to a plant's cell wall.

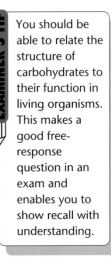

EXAMINER'S TIP

You should be able to relate the structure of carbohydrates to their function in living organisms. This makes a good free-response question in an exam and enables you to show recall with understanding.

Proteins

In any cell, proteins form:
◆ part of its membranes and membrane-bound organelles
◆ a major constituent of the fluid cytoplasm that surrounds its organelles
◆ its enzymes.

PROTEIN MONOMERS: AMINO ACIDS

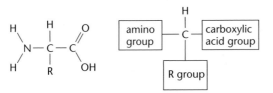

FIGURE 3.8 An amino acid has an amino group (NH_2) and a carboxyl group (COOH) attached to a single carbon atom. Different amino acids have different R groups.

Figure 3.8 shows the generalised structure of an amino acid. It has a central carbon atom attached to:
◆ an amino group (NH_2)
◆ a carboxyl group (COOH)
◆ an R group which is different in each of the 20 amino acids that commonly occur in natural proteins.

Two amino acids can join together in a condensation reaction. Figure 3.9 shows this condensation and the formation of a peptide bond that holds amino acids together.

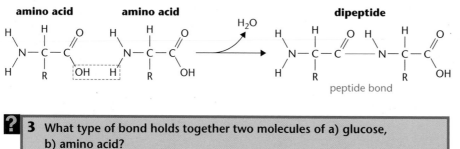

FIGURE 3.9 Two amino acids form a dipeptide during a condensation reaction. A peptide bond holds the amino acids together.

? 3 What type of bond holds together two molecules of a) glucose, b) amino acid?

? 4 Which part of the amino acid shown in Figure 3.8 makes the amino acids glycine and cysteine different from each other?

PROTEIN POLYMERS: POLYPEPTIDES AND PROTEINS

Large numbers of amino acids can join together by condensation reactions to form polypeptides and proteins.

◆ Polypeptides are long chains of amino acids joined together by peptide bonds.
◆ Proteins contain one or more polypeptides that are folded into a complex shape.

Protein structure has four aspects.

1 Primary structure – the sequence of amino acids (see Figure 3.10).
2 Secondary structure – a polypeptide forms a helix or several polypeptides form a folded sheet (see Figure 3.11). In both cases, hydrogen bonds hold the shapes in place.
3 Tertiary structure – the secondary structure folds to give a complex, three-dimensional shape (see Figure 3.12).

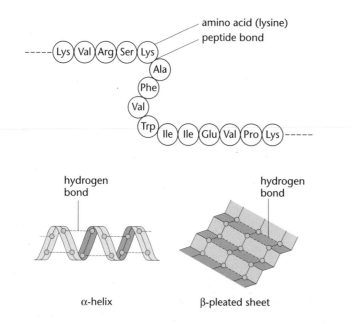

FIGURE 3.10 The sequence of amino acids gives a polypeptide its primary structure.

FIGURE 3.11 An individual polypeptide can form an α-helix or several can form a folded sheet. In both cases, hydrogen bonds hold the shape in place.

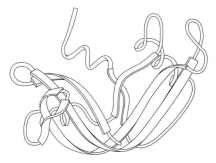

FIGURE 3.12 This diagram shows the tertiary structure of a molecule of ribonuclease, a protein. Disulphide bridges between molecules of the sulphur-containing amino acid cysteine help to keep this shape. The spiral parts of the diagram each contain an α-helix and the flat parts each contain pleated sheets.

Such proteins are often called globular proteins. Their shape is crucial to their function, which might be:

◆ an enzyme with an active site that must complement the shape of its substrate
◆ a receptor protein with a shape that must complement its target hormone
◆ a 'recognition' protein on the outside of lymphocytes that 'recognises' antigens on an invading bacterial cell.

The shape of globular proteins is held in place by:

◆ numerous, but weak, hydrogen bonds between the R groups of adjacent amino acids in the chain
◆ occasional, but strong, disulphide bridges between sulphur-containing amino acids, such as cysteine.

Heating a protein causes these bonds to break, so that the protein loses its shape and its properties.

4 Quaternary structure – two or more polypeptide chains combine to produce a functional protein. For example, a single haemoglobin molecule is made from four polypeptide chains (see Figure 3.13).

EXAMINER'S TIP

When asked about the denaturation of proteins, remember it is the hydrogen bonds and disulphide bridges that are broken. You will not gain credit if you write that the peptide bonds are broken.

EXAMINER'S TIP

You will not be expected to recall the structure of any named amino acid or polypeptide. However, you might be asked to interpret diagrams of amino acids or polypeptides.

Haemoglobin
one of the four polypeptide chains that make up a molecule of haemoglobin

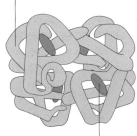

FIGURE 3.13 Haemoglobin is the red oxygen-carrying pigment found in human blood. It is a globular protein formed from four polypeptide chains each attached to an iron-containing haem group. It is the haem groups that combine with oxygen.

A haem group:
each polypeptide chain is attached to a haem group. This group is important in transporting oxygen

5 Enzymes are proteins. Suggest why enzymes lose their properties when heated.

Lipids: triglycerides and phospholipids

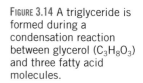

6 Is a triglyceride a polymer? Explain your answer.

A triglyceride consists of:
- a molecule of glycerol ($C_3H_8O_3$) combined with
- three fatty acids (long chains of carbons, simplified as RCOOH).

The way in which these molecules combine together during a condensation reaction is shown in Figure 3.14.

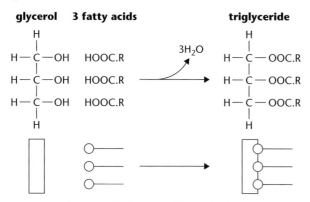

FIGURE 3.14 A triglyceride is formed during a condensation reaction between glycerol ($C_3H_8O_3$) and three fatty acid molecules.

There are two types of fatty acid, shown in Figure 3.15.
- Unsaturated fatty acids have double bonds between some of the carbon atoms along their length. Such fatty acids have low melting points and the unsaturated triglycerides formed from them are liquid at room temperature, e.g. plant oils.
- Saturated fatty acids have no double bonds between carbon atoms along their length. Such fatty acids have higher melting points and the saturated triglycerides formed from them are solid at room temperature, e.g. animal fats.

saturated fatty acid

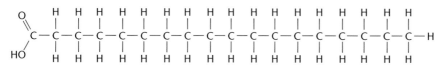

unsaturated fatty acid

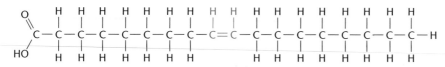

FIGURE 3.15 In saturated fatty acids, the carbon atoms are joined together by single covalent bonds. In unsaturated fatty acids, at least some of the carbon atoms are joined together by double covalent bonds.

A phospholipid is shown in Figure 3.16. It is similar to a triglyceride except that a phosphate group replaces one of the fatty acids. As a result, a phospholipid has two ends.
- A 'head' containing the phosphate group. This has an uneven charge that makes the head able to mix with water (hydrophilic).
- A 'tail' containing two fatty acid chains. They don't have an uneven charge, making them unable to mix with water (hydrophobic).

As a result of their hydrophilic head and hydrophobic tail, phospholipids will arrange themselves in a double layer when mixed with water (see membrane structure in Chapter 2).

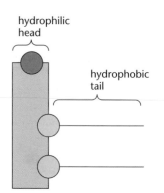

hydrophilic head

hydrophobic tail

FIGURE 3.16 A phospholipid molecule is like a triglyceride in which one of the fatty acids has been replaced by a phosphate group. The phosphate group gives a phospholipid a hydrophilic 'head' and a hydrophobic 'tail'.

Practical skills

BIOCHEMICAL TESTS

Table 3.2 summarises the simple biochemical tests that you should be able to recall and interpret.

Compound to be tested for	Description of test	Positive result
Reducing sugar	Add Benedict's solution to test solution. Boil briefly in a water bath.	Colour of solution turns from blue to brick-red.
Non-reducing sugar	As above. If no brick-red colour appears, boil fresh sample of test solution with dilute hydrochloric acid (this breaks the non-reducing sugar down into its constituent monomers). Cool and add enough sodium hydrogencarbonate to neutralise the acid. Then add Benedict's solution to test solution. Boil briefly in a water bath.	 Colour of solution turns from blue to brick-red.
Starch	Add a few drops of iodine/potassium iodide solution to the test solution.	Colour of iodine solution turns from yellow to blue-black.
Protein	Add a few drops of biuret reagent to the test solution.	Colour of biuret solution turns from blue to lilac.
Triglyceride (lipid)	Mix test material with water and shake. Emulsion forms that separates when left to stand. Mix fresh test material with dilute alcohol solution.	 Emulsion formed which does not separate when left to stand.

TABLE 3.2 Tests for carbohydrates, proteins and triglycerides

EXAMINER'S TIP

Candidates commonly make two mistakes when referring to biochemical tests in unit tests. Make sure you refer correctly to:
- when solutions should be heated (only in the Benedict's test)
- the actual appearance, or colour change, of the positive result.

?

7 When heated with a sucrose solution, Benedict's solution would remain blue.
a) How could you demonstrate that sucrose is a non-reducing sugar?
b) From your knowledge of the structure of sucrose, explain why Benedict's solution would turn brick-red after your treatment of the sucrose.

CHROMATOGRAPHY

Chromatography is a technique that is used to separate mixtures of compounds and to identify the compounds separated from the mixture. The method described in Table 3.3 uses paper as the absorbent base and is the one you are most likely to use in your practical work. Several other absorbent bases could be used.

Procedure	Explanation
Grind up the tissue containing these substances using a pestle and mortar and an appropriate solvent.	Extracts the substances from the tissue.
Rule a pencil line near the bottom of the piece of chromatography paper. This is the origin. Draw pencil crosses on this line where you will apply spots of the mixture and of any other reference substance.	Once you have run the chromatogram, you will know where the spots of reagents started. Use a pencil since ink might dissolve in the solvent you will use.
Use a fine pipette to form a small spot of mixture on the origin line. Dry the spot before adding another spot of mixture, then repeat.	Drying ensures the spot remains small. Adding more drops ensures the final spot is concentrated.
Suspend the chromatography paper in a jar containing the solvent. Ensure that the paper is in the solvent but that the level of the solvent is below the origin line.	This ensures that the solvent will take up the spots you have made as it soaks up the paper.
Cover the jar and leave it undisturbed until the solvent nears the top of the chromatography paper.	This ensures maximum separation of the components in each spot.
Remove the paper from the jar and mark the position of the solvent front with a pencil line.	You need to know how far from the origin line the solvent has travelled.
If you are separating coloured compounds, you will see several spots of colour on the chromatogram (see Figure 3.17). If not, after drying, spray the chromatography paper with a locating agent.	Spraying with a locating agent colours the separated compounds, enabling you to see them.
If chromatography has not separated the compounds very well, turn the chromatogram through 90° and run it again using a different solvent. This is called two-way chromatography and is shown in Figure 3.18.	Running the chromatogram again separates the compounds better than one-way chromatography.

TABLE 3.3 The steps involved in paper chromatography

? 8 Why must you use a **pencil** to draw the lines on a chromatogram and not a pen?

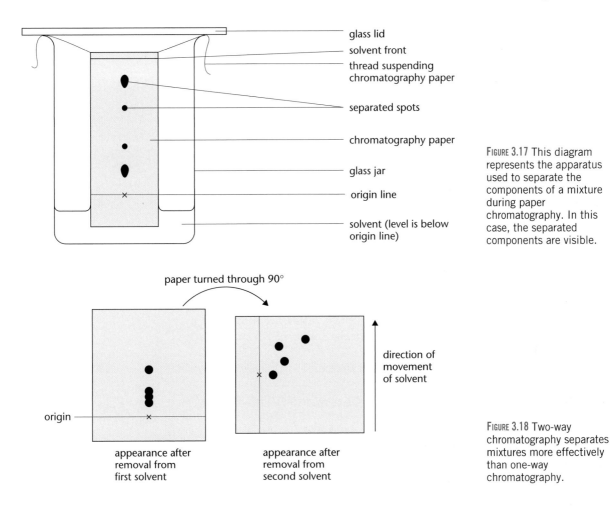

FIGURE 3.17 This diagram represents the apparatus used to separate the components of a mixture during paper chromatography. In this case, the separated components are visible.

FIGURE 3.18 Two-way chromatography separates mixtures more effectively than one-way chromatography.

The position of the separated compounds on the chromatography paper allows them to be identified. To do this, you calculate the **Rf value** using the formula:

$$\text{Rf value} = \frac{\text{Distance moved by substance}}{\text{Distance moved by solvent front}}$$

For any given solvent, the Rf value for each compound is constant. This allows you to identify the compounds by comparing their Rf values with those in a reference book. Alternatively, you can put spots of reference compounds on the origin of the chromatogram and compare the distance they move with spots from the unknown mixture.

9 Calculate the Rf value of the compound that has moved furthest from the origin in Figure 3.17.

EXAMINER'S TIP

Make sure that you can calculate Rf values from drawings of chromatograms. Examiners often use this as a test of Assessment Objective 2.

When measuring how far a separated spot has travelled from the origin, always make it clear in your explanation whether you measured to the centre of a spot or to its front edge.

WORKED EXAM QUESTION

1   a)   (i) How many molecules are produced when a triglyceride molecule is completely hydrolysed? *(1 mark)*

Three.

This candidate seems to have locked on to the 'tri' part of 'triglyceride' and has not recalled that a triglyceride is made of one molecule of glycerol linked to three fatty acid molecules (i.e. four molecules in all).

(ii) Many large biological molecules are polymers. Explain why triglycerides are **not** polymers. *(1 mark)*

Because this molecule is not a long chain. Molecules of starch are made from amylose and amylopectinose and are much longer than triglycerides.

Apart from making an error in recalling the name of amylopectin, this candidate has missed the point of the question. The correct answer is that triglycerides are not made of identical units/monomers.

b)   Molecules can be represented in different ways. **Figure 1** shows a model of a fatty acid. It shows the different atoms that make up the molecule.

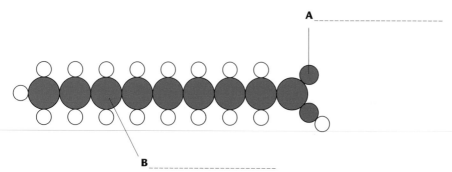

Figure 1

(i) Complete the diagram by naming the atoms labelled **A** and **B**.   *(2 marks)*

A   *This is an oxygen atom.*
B   *This is a carbon atom.*

The candidate correctly identified the labelled atoms. Answers of 'oxygen' and 'carbon' or simply 'O' and 'C' would have gained full marks.

(ii) This molecule is a saturated fatty acid. Explain the meaning of *saturated*.

(*1 mark*)

> *This means that the fatty acid is saturated and cannot take*
> *any more.*

It is not acceptable to explain a meaning simply by repeating the key word. This candidate has failed to recall the technical meaning of 'saturated' and has, instead, given a general meaning ('cannot take any more'). A mark would have been awarded for 'it has no double bonds between its carbon atoms'.

c) A droplet of phospholipid was put into a large dish of water. The drop had a volume of 1 mm³. It spread out to form a thin film on the surface of the water which covered an area of 400 000 mm². **Figure 2** shows the appearance of the surface film formed by the phospholipid molecules.

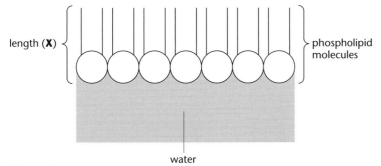

Figure 2

(i) Calculate the length (**X**) of a single phospholipid molecule. Show your working. (*2 marks*)

> *Measured length of X is 3.5 cm Magnification is 400 000 times*
> *Therefore actual length of X is 3.5 × 400 000 = 1 400 000*

The candidate has shown his working, which is good, but has made a number of errors. One error commonly made in calculation questions is that he has failed to give units to his numerical answer. The candidate has made a more fundamental error in interpreting the question. This question does not ask for a measurement to be made and then scaled. Instead, the question gives you a volume and a surface area and expects you to *know* that volume is calculated as surface area × length. In this case, the required length is calculated as volume divided by surface area (1 ÷ 400 000), which is 0.0000025 mm (also expressed correctly as 2.5×10^{-6} mm). The correct answer on its own would gain two marks. If you had correctly shown how you were making the calculation but then made an arithmetic error, you would still gain one mark for showing the correct working.

(ii) Explain what causes the phospholipid molecules to be arranged in the way shown in **Figure 2**. (*2 marks*)

> *The tails are hydrophobic and so will not mix with the water.*

This statement is correct and gains one mark. To gain the second mark, the candidate needs to explain that the heads are hydrophilic and so will mix with water. Always use the mark scheme to judge how many points you should make in your answer.

(*AQA 2003*)

EXAMINATION QUESTION

1   A tripeptide is made of three amino acids. The diagram shows the molecular structure of a tripeptide.

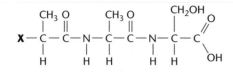

a) (i)   Give the formula of the chemical group at position **X** on the molecule.

(1 mark)

(ii) Give **one** piece of evidence from the diagram that this molecule is made up from three amino acids. (1 mark)

b) This tripeptide was broken down into its amino acids. These were separated and identified using chromatography. The diagram shows the resulting chromatogram.

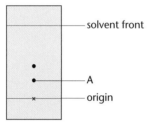

(i)   Mark the diagram with a line to show where the solvent should come to when the apparatus is set up. (1 mark)

(ii) The tripeptide was completely broken down into its amino acids but there are only two spots on the chromatogram. Explain why. (1 mark)

(iii) Give the formula which would allow you to calculate the Rf value of amino acid **A** from the chromatogram. (1 mark)

(iv) To identify an amino acid, it is necessary to calculate Rf values rather than measure only the distance moved by the spot. Explain why.

(2 marks)

(AQA 2001)

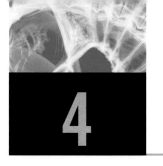

4 Enzymes

◆ All enzymes are proteins with a specific tertiary shape.

◆ Enzymes act as catalysts – they speed up the rate of reaction.
  They are highly specific, being used in one reaction only.

◆ Enzyme molecules and substrate molecules in a solution, as in a cell, are constantly moving around. They must collide with each other before they will react. Collisions are random. The more collisions that take place, the greater the rate of reaction.

◆ Enzymes remain unchanged at the end of a reaction and are therefore able to be used again.

◆ Enzymes are affected by temperature, pH, inhibitors, concentration of enzyme and concentration of substrate.

? **1** Suggest **one** advantage of enzymes being protein molecules.

Activation energy

◆ In a reaction, energy is needed to break chemical bonds so that new ones can be formed.

◆ The energy necessary to break these bonds is the activation energy.

◆ Enzymes speed up the rate of reaction by lowering the activation energy necessary to start the reaction (see Figure 4.1).

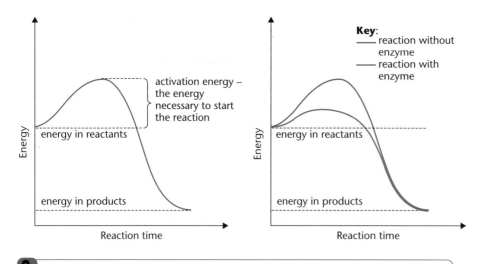

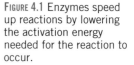

activation energy –
the energy
necessary to start
the reaction

energy in reactants

energy in products

Reaction time

Energy

energy in reactants

energy in products

Reaction time

Energy

Key:
—— reaction without
enzyme
—— reaction with
enzyme

Figure 4.1 Enzymes speed
up reactions by lowering
the activation energy
needed for the reaction to
occur.

EXAMINER'S TIP

When drawing a curve to show activation energy, remember that the curve
must start and finish at the same level as it would be without an enzyme. The
only difference is that the activation energy is much less with an enzyme.

How enzymes work

Enzymes are proteins with a specific tertiary structure. A pocket within the protein is
known as the active site. The active site is part of the enzyme into which a substrate fits,
forming an enzyme–substrate complex. The shape of the active site differs between
enzymes, so each enzyme is specific to a particular substrate. When the complex has
formed, a reaction takes place and the products are released from the active site leaving
the enzyme free to combine with another substrate molecule.

? 2 Explain how an enzyme reduces the activation energy.

one of the amino acids which
helps to make the active site
of this enzyme a specific shape

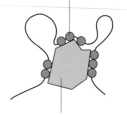

the substrate is a
complementary shape to that
of the active site of the
enzyme

Figure 4.2 Each enzyme has an active site of a specific shape
to fit a particular substrate.

Two models have been put forward to explain how enzymes work:

1 the lock and key hypothesis
2 the induced fit hypothesis.

1 LOCK AND KEY HYPOTHESIS

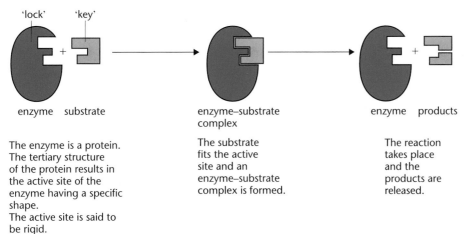

'lock' 'key'

enzyme substrate

enzyme–substrate complex

enzyme products

The enzyme is a protein. The tertiary structure of the protein results in the active site of the enzyme having a specific shape.
The active site is said to be rigid.

The substrate fits the active site and an enzyme–substrate complex is formed.

The reaction takes place and the products are released.

FIGURE 4.3 The lock and key hypothesis of enzyme action

2 INDUCED FIT

The induced fit hypothesis is very similar to the lock and key hypothesis. The only difference is that the active site moulds round the substrate. The enzyme is said to be flexible.

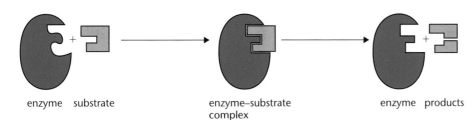

enzyme substrate

enzyme–substrate complex

enzyme products

The active site is flexible and takes on the shape of the substrate.

FIGURE 4.4 The induced fit hypothesis of enzyme action

> **EXAMINER'S TIP**
>
> It is important that you do not say that the substrate has the same shape as the active site. It is more precise to say that the shape of the substrate is complementary to that of the active site.

An enzyme-controlled reaction depends on substrate molecules fitting into the active site of the enzyme. Environmental factors which affect the ability of the active site to combine with substrate will alter the rate of reaction. These factors are:

◆ temperature
◆ pH
◆ inhibitors
◆ concentration of substrate
◆ concentration of enzyme.

THE EFFECT OF TEMPERATURE

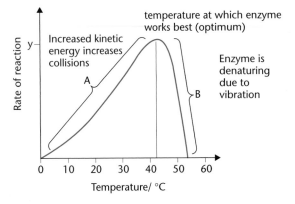

FIGURE 4.5 Increasing the temperature increases the rate of a reaction only to the point at which an increase in the number of enzyme molecules denatured balances out the increased rate of collisions.

Region A on the graph	Region B on the graph
Description As the temperature increases from 0 °C to 40 °C, the rate of reaction increases. At 40 °C, it reaches its maximum rate of reaction (y). **Explanation** ◆ Increasing the temperature provides more heat energy. ◆ This gives both the enzyme and the substrate more kinetic energy. ◆ Both enzyme and substrate move faster. ◆ This results in more collisions. ◆ More enzyme–substrate complexes are formed. ◆ The rate of reaction is increased.	**Description** As the temperature increases above 40 °C, the rate of reaction decreases until at 55 °C the rate of reaction is zero. **Explanation** ◆ High temperatures result in more kinetic energy being given to the enzyme and the substrate. ◆ The enzyme molecules vibrate more as the kinetic energy increases. ◆ The bonds holding the tertiary structure of the protein break (usually the hydrogen bonds). ◆ This changes the shape of the enzyme. ◆ The active site changes shape. ◆ The substrate can no longer fit. ◆ There is less/no reaction occurring. ◆ We say that the enzyme has been denatured.

TABLE 4.1 The effect of temperature on an enzyme-controlled reaction

EXAMINER'S TIP

◆ Not all enzymes have an optimum temperature of 40 °C (or body temperature). Make sure you read the information given in the question or look carefully at the data given in the graph.
◆ Read the question carefully. Sometimes the question asks about what happens at low temperatures.
◆ When giving a description of the changes shown in a graph, always use the information on the axes. **Never** say only that it is increasing or decreasing.
◆ When giving an explanation, give the detailed biology of why it is happening.
◆ Try to give the whole story when answering a question.
◆ If given the name of the enzyme or substrate or both, use them in your answer.
◆ Never say that enzymes are 'killed'.
◆ Denaturation is only one word and therefore worth only 1 mark. Usually a question will have 2 or 3 marks and so some detail of what is happening during denaturation is required.

THE EFFECT OF pH

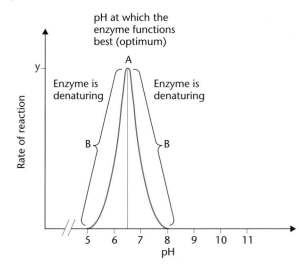

FIGURE 4.6 Each enzyme has a specific optimum pH at which it works best. For this enzyme, the optimum pH is 6.5. At this pH, the rate of reaction is fastest.

Region A on the graph	Region B on the graph
Description At a pH of 6.5, there is a maximum rate of reaction (y) or pH 6.5 is the optimum.	**Description** If you change the pH away from its optimum (either more or less acidic), the rate of reaction decreases. At a pH of beyond 5 or 8 in both directions there is no reaction.
Explanation ◆ A solution at the optimum pH has the correct number of hydrogen ions and does not disrupt the ionic bonding for this enzyme. ◆ There is no change in the shape of the enzyme. ◆ There is no change in the shape of the active site. ◆ The substrate can fit. ◆ The enzyme is able to form enzyme–substrate complexes. ◆ Rate of reaction is at its maximum.	**Explanation** ◆ The change in pH has disrupted the ionic bonding in the tertiary structure of the protein. ◆ This changes the shape of the enzyme. ◆ This changes the shape of the active site. ◆ The substrate cannot fit. ◆ Enzyme–substrate complexes are not formed. ◆ The rate of reaction is less/none.

Remember
◆ Enzymes usually function in a very narrow pH range.
◆ Altering the pH beyond the narrow range in which an enzyme functions results in denaturation.
◆ A change in pH alters the number of hydrogen ions (H^+) in the surrounding solution.
◆ More acidic means more hydrogen ions, more alkaline means fewer hydrogen ions.
◆ Any change from the optimum pH disrupts the ionic bonding which is holding the tertiary structure of the protein in the correct shape.

TABLE 4.2 The effect of pH on an enzyme-controlled reaction

THE EFFECT OF INHIBITORS

All inhibitors slow down the rate of an enzyme-controlled reaction. There are two types of inhibitors:

1 competitive inhibitors
2 non-competitive inhibitors.

1 Competitive inhibitors

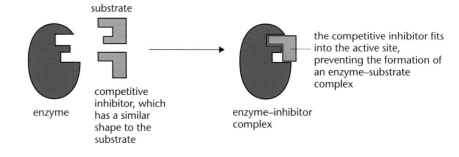

FIGURE 4.7 The competitive inhibitor has a similar shape to the substrate and blocks the active site.

◆ The competitive inhibitor has a similar shape to the substrate.
◆ The inhibitor fits into the active site, forming an enzyme–inhibitor complex.
◆ This stops the substrate from entering the active site of the enzyme.
◆ Fewer enzyme–substrate complexes are formed.
◆ The rate of reaction thus decreases.

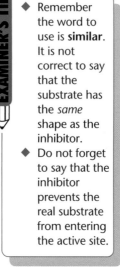

EXAMINER'S TIP
◆ Remember the word to use is **similar**. It is not correct to say that the substrate has the *same* shape as the inhibitor.
◆ Do not forget to say that the inhibitor prevents the real substrate from entering the active site.

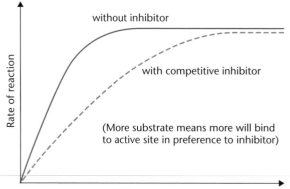

FIGURE 4.8 Up to a point, increasing the substrate concentration increases the rate of reaction, even when a competitive inhibitor is present.

◆ The extent to which a competitive inhibitor reduces enzyme activity depends on the relative number of substrate molecules.
◆ Remember that an enzyme collides with a substrate or an inhibitor by chance.
◆ If there are twice as many substrate molecules, there will be more chance that the enzyme will meet the substrate rather than the inhibitor. If you have many substrate molecules (i.e. a concentrated substrate solution), there will be little chance that the enzyme will meet the inhibitor.

2 Non-competitive inhibitors

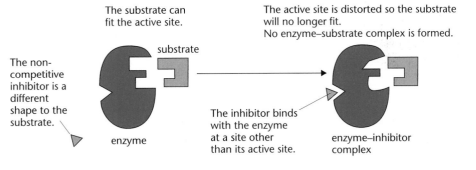

The non-competitive inhibitor is a different shape to the substrate.

The substrate can fit the active site.

substrate

enzyme

The inhibitor binds with the enzyme at a site other than its active site.

The active site is distorted so the substrate will no longer fit.
No enzyme–substrate complex is formed.

enzyme–inhibitor complex

FIGURE 4.9 The non-competitive inhibitor binds to the enzyme and changes the shape of the active site. The substrate molecule no longer fits.

◆ The non-competitive inhibitor is a different shape to the substrate.

◆ It will never fit into the active site.

◆ The inhibitor fits into a site other than the active site.

◆ This distorts the active site.

◆ The substrate can no longer fit into the active site.

◆ No enzyme–substrate complex is formed.

◆ The rate of reaction thus decreases.

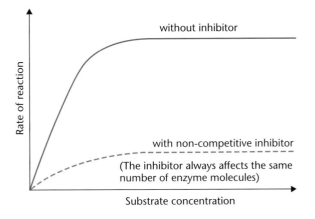

without inhibitor

Rate of reaction

with non-competitive inhibitor
(The inhibitor always affects the same number of enzyme molecules)

Substrate concentration

FIGURE 4.10 The effect of a non-competitive inhibitor on the rate of an enzyme-controlled reaction.

The effect of non-competitive inhibitors is largely independent of the substrate concentration. As long as there are enough inhibitor molecules to bind with all the enzymes, there will be 100% inhibition. If there are enough inhibitor molecules to bind with 80% of the enzymes, there will only be 80% inhibition (as shown in Figure 4.10).

3 **The lock and key model relies on the fact that the enzyme is a rigid structure. Why is this not a good model to explain how a non-competitive inhibitor functions?**

THE EFFECT OF ENZYME CONCENTRATION

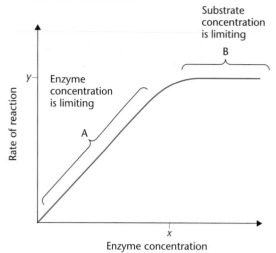

FIGURE 4.11 Increasing the enzyme concentration increases the rate of reaction up to the point where the substrate concentration becomes the limiting factor.

Region A on the graph	Region B on the graph
Description	**Description**
As the enzyme concentration increases, so does the rate of reaction. At an enzyme concentration of x, the rate of reaction is at its maximum rate y.	As the concentration of enzyme increases above a concentration of x, the rate of reaction remains constant at its maximum rate of reaction, y.
Explanation	**Explanation**
◆ With more enzyme molecules, there are more active sites available for the substrate to fit. ◆ More enzyme–substrate complexes are formed. ◆ More products are formed. ◆ The rate of reaction is increased. ◆ The limiting factor is enzyme concentration.	◆ All the substrates have formed enzyme–substrate complexes. ◆ The enzyme concentration is no longer the limiting factor. ◆ The limiting factor is substrate concentration.

TABLE 4.3 The effect of enzyme concentration on an enzyme-controlled reaction

? 4 Why are enzymes only needed in small concentrations?

THE EFFECT OF SUBSTRATE CONCENTRATION

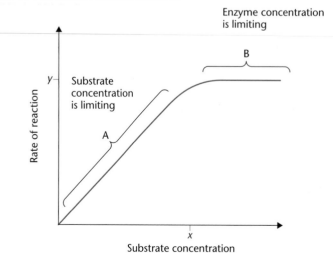

FIGURE 4.12 Increasing the substrate concentration increases the rate of reaction up to the point where the enzyme concentration becomes the limiting factor.

Region A on the graph	Region B on the graph
Description As the substrate concentration increases, the rate of reaction increases. At a substrate concentration of x, there is a maximum rate of reaction, y. **Explanation** ◆ The more substrate molecules there are, the greater the chance of collision with the enzyme. ◆ More collisions so more enzyme–substrate complexes are formed. ◆ Increased rate of reaction. ◆ Substrate concentration is limiting.	**Description** As the substrate concentration increases above value x, the rate of reaction remains constant at its maximum, y. **Explanation** ◆ Increasing the substrate concentration makes no difference to the rate of reaction. ◆ All the enzyme's active sites are occupied. ◆ The enzymes are working at their maximum turnover rate. ◆ Enzyme concentration is limiting.

TABLE 4.4 The effect of substrate concentration on an enzyme-controlled reaction

EXAMINER'S TIP
◆ The limiting factor in region A of a graph of this shape is always what is given on the x-axis.
◆ When asked to explain why a factor is 'limiting', the only answer is that if all other conditions were kept the same and this factor was increased, the rate of reaction would increase.

A note on coursework

Enzymes are often taught practically and, as such, they make a good topic for the practical assessment. Many different practicals can be attempted so it is important that you understand the principles. These principles can also be tested in a written paper.
◆ Whatever the independent variable, you must remember to keep all the other variables that can affect enzymes constant. For example, if you are going to change pH (the independent variable), then temperature, substrate concentration and enzyme concentration must be kept constant.
◆ The independent variable must have a good range. For example, five different pHs, such as pH 2, 4, 6, 8 and 10.
◆ You must give enough detail of how you are going to keep the other variables constant. For example, the temperature will be kept at 40 °C by using a thermostatically controlled water bath.
◆ Remember that the enzyme and the substrate must both be brought up to temperature *separately* before they are mixed.
◆ If a rate of reaction is to be measured, make sure that a sensible time scale is chosen.

EXAMINER'S TIP
Do not forget that as an enzyme-controlled reaction progresses, the number of products produced will be different at the end than at the start. So if you are going to compare two different reactions, you must always compare the initial rate.

WORKED EXAM QUESTION

1   a)   When heated, hydrogen peroxide breaks down to water and oxygen.

$$2H_2O_2 \longrightarrow 2H_2O + O_2$$

The graph shows the energy changes which take place during this reaction.

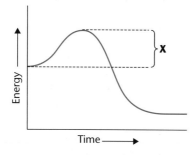

(i)   What is represented by the part of the curve labelled X?   *(1 mark)*

Activation energy

In all chemical reactions, the energy must at first be increased to allow the molecules to react together. At AS level there are a number of facts that you just have to know. In this question the name of the energy change was required. In another question the name of the process may be given and you will be expected to explain what it means, see pages 39–40. Be prepared for either.

(ii)   This reaction also takes place in many living cells. Here it is catalysed by the enzyme catalase. Sketch a curve on the graph to show the energy changes which take place when the enzyme is present.   *(2 marks)*

[Curve drawn lower than X; but falling below the lowest level of the original curve]

Enzymes lower the activation energy and thus the first part of this curve is correct. However enzymes do not alter the products formed, therefore the energy level of the products would be the same. The curve will not fall below the original as suggested by this candidate.

(iii)   When the reaction with catalase is carried out in a test tube, the test tube feels warm at the end of the reaction. Use the graph to explain why.   *(2 marks)*

The test tube feels warm at the end of the reaction because heat is being produced by the reaction. The graph shows this really clearly.

There are two marks available here, but recognising that the test tube will get hot if heat is given off is not quite enough for either.

The candidate has realised that he has to mention the graph but simply to say 'the graph shows this', without clearly explaining how, is not good enough. You need to understand what the graph is showing. It represents the amount of energy in the system. As the level of energy shown by the graph is lower at the end of the reaction, then there must be less energy in the products than in the substrates/raw materials. That is one of the marking points. That energy must be released in the form of heat is the other.

b) The turnover number of an enzyme is the number of substrate molecules converted to product per second. The maximum turnover number of catalase is 200 000 molecules per second. Explain why the turnover number falls as the temperature gets lower. *(2 marks)*

> *As the temperature falls the molecules have less kinetic energy and so they move less. Fewer substrate molecules collide with enzyme molecules and fewer products are formed.*

This is a really clear and logical answer. Always try to think of enzyme reactions occurring in stages. Enzyme and substrate meet; the active site is complementary to the substrate; an enzyme–substrate complex is formed; products are made and released. Anything that encourages the collision of enzyme and substrate (like increased temperature) will increase turnover rate; anything that alters the shape of the enzyme (like high temperature, which causes vibration of the enzyme, or changing pH) will decrease the turnover rate.

(AQA 2002)

EXAMINATION QUESTION

1   The total amount of product formed in an enzyme-controlled reaction was investigated at two different temperatures, 55 °C and 65 °C. The results are shown in the graph.

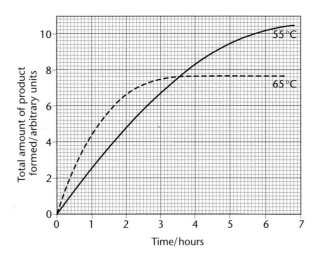

a) (i) Explain how you would calculate the rate of the reaction at 55 °C over the first 2 hours of the investigation.   (1 mark)

   (ii) Explain why the initial rate of this reaction was faster at 65 °C than it was at 55 °C.   (3 marks)

b) Use your knowledge of enzymes to explain the difference in the two curves between 4 and 6 hours.   (1 mark)

c) In this investigation, the enzyme and its substrate were mixed in a buffer solution. What was the purpose of the buffer solution?   (1 mark)

(AQA 2001)

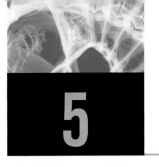

5 Gas exchange in humans

After revising this topic, you should be able to:

▶ describe the gross structure of the human gas exchange system

▶ use Fick's law to show how the alveolar epithelium provides an efficient gas exchange surface

▶ interpret data to show an understanding of the exchange of gases within the human gas exchange system

▶ show an understanding of the mechanism of ventilation and the nervous control of its rhythm

▶ show a qualitative and quantitative understanding of the effect of exercise on pulmonary ventilation, tidal volume and breathing rate.

Respiration and gas exchange

Most reactions that occur in cells use energy. Since the most common source of this energy is the breakdown of adenosine triphosphate (ATP), cells must have a store of ATP if they are to remain active. Cells produce ATP in their mitochondria in a process called aerobic respiration. You will learn about respiration in your A2 course. For the moment, you need only know that aerobic respiration:

- uses organic molecules, usually glucose, as the energy source
- uses oxygen – as a result, the oxygen level is depleted in all respiring cells
- produces carbon dioxide – as a result, carbon dioxide accumulates in all respiring cells
- produces ATP.

Respiring cells can remain active only if they gain more oxygen and lose carbon dioxide. In humans, both oxygen and carbon dioxide are carried in the blood between respiring cells and the gas exchange surface of the lungs. You must be familiar with the process of gas exchange in the lungs.

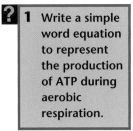 **1** Write a simple word equation to represent the production of ATP during aerobic respiration.

Structure of the lungs

GROSS STRUCTURE

The lungs lie in the thoracic cavity (or chest cavity), surrounded by:

- twelve pairs of ribs, with intercostal muscles between them
- the diaphragm.

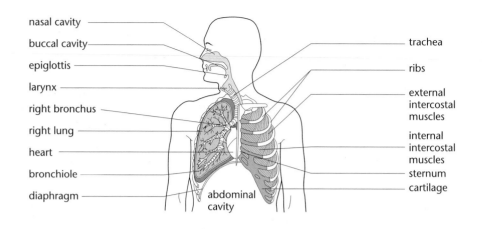

Figure 5.1 The gross structure of the human gas exchange system

Figure 5.1 shows the internal structure of the lungs and how they connect to the atmosphere. Notice the following features of the gas exchange system in Figure 5.1.

◆ The nasal cavity (for breathing) and the buccal cavity (part of the digestive system) are separated from each other.

◆ The epiglottis is able to close the entrance from the pharynx into the trachea, preventing food entering the gas exchange system.

◆ The trachea, bronchi and larger bronchioles have rings of cartilage. Since each of these tubes has a large lumen with a sticky inner surface, the cartilage prevents them closing up.

◆ The epithelium lining the trachea and the bronchi secretes mucus and is coated with tiny, hair-like cilia (see Figure 5.2). The mucus traps particles in the inhaled air, including bacteria. Beating of the cilia moves mucus up the trachea into the pharynx, where it is swallowed.

◆ The bronchi divide into a series of bronchioles which branch to all parts of the lungs.

◆ Most of the lung tissue is filled with alveoli.

2 Put these terms into the correct sequence, starting with the part nearest the atmosphere: alveolus, bronchus, bronchiole, epiglottis, pharynx, trachea.

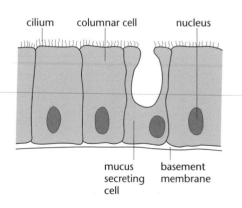

Figure 5.2 A ciliated epithelium lines the trachea and the bronchi. The movement of the cilia moves mucus, with trapped particles, back up to the pharynx where it is swallowed.

3 Substances in cigarette smoke stop the action of cilia in the trachea, bronchi and bronchioles.
a) Explain why smoking leads to coughing.
b) Explain why smoking might increase the risk of lung infections.

ALVEOLI

Each of the smallest bronchioles ends in a group of air sacs called alveoli. The walls of the alveoli form the gas exchange surface through which:

◆ oxygen diffuses from the inhaled air into the blood
◆ carbon dioxide diffuses from the blood into the air to be exhaled.

Alveoli are well adapted to their role as gas exchange surfaces.

◆ Because there are so many of them, the alveoli form a large surface area.
◆ The blood capillary network around the alveoli brings carbon dioxide to them and takes away oxygen to the respiring tissues (see Figure 5.3).
◆ Each alveolus is one cell thick, i.e. is a simple epithelium. The cells in this epithelium are squamous cells, i.e. they are very thin (see Figure 5.4).
◆ The air space in Figure 5.4 is separated from the blood plasma in the capillary by only two layers of squamous cells: this is a very short diffusion path.

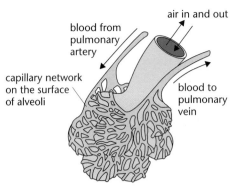

FIGURE 5.3 A bronchiole ends in a group of alveoli. A dense capillary network surrounds each group of alveoli.

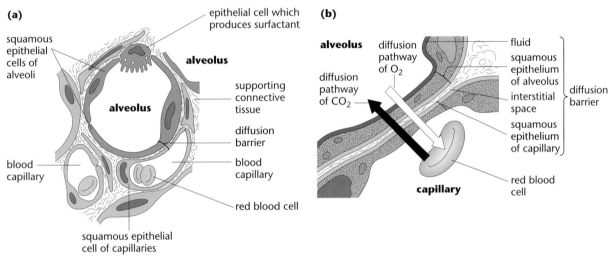

FIGURE 5.4 (a) A drawing made from an electron micrograph showing several alveoli and capillaries.
(b) A simplification of part of the drawing in (a) to show the diffusion pathways of oxygen and carbon dioxide.

Gas exchange across the alveoli

Figure 5.4(b) represents the diffusion pathways of oxygen and carbon dioxide. In both cases, the only barrier to diffusion is formed by two layers of squamous epithelium separated by a thin film of fluid.

Air	Partial pressure/ kPa			
	Nitrogen	Oxygen	Carbon dioxide	Water
Inspired	79.6	21.2	0.04	0.5
Alveolar	75.9	13.9	5.3	6.3
Expired	75.5	16.0	3.6	6.3

TABLE 5.1 Standard values for respiratory gases in inspired, alveolar and expired air

? 4 a) Write an equation to represent Fick's law.
b) Use Fick's law to show how the alveoli are adapted as an efficient exchange surface.

Table 5.1 shows the effect of gas exchange in the alveoli. Expired air has a lower concentration of oxygen and a higher concentration of carbon dioxide than inspired air. Note other information in Table 5.1.

- The concentration of oxygen in expired air is about 75% that of inspired air. The human gas exchange system is not very efficient at extracting oxygen from air.

- The composition of expired air is different from that of alveolar air. This is because expired air contains alveolar air mixed with unchanged inspired air that was held in the trachea, bronchi and larger bronchioles.

- The amount of water vapour in the air increases.

- There is a small variation in the partial pressure of nitrogen throughout the respiratory system. This is because the nitrogen is diluted by water vapour and carbon dioxide from the lungs, not because it diffuses between alveoli and blood.

---

5 Explain why carbon dioxide diffuses from plasma surrounding the alveoli into the air within the alveoli.

6 a) Explain why the composition of inspired air that is held in the trachea, bronchi and bronchioles does not change.
   b) Explain why the amount of water vapour in expired air is greater than that in inspired air.

EXAMINER'S TIP

Make sure you practise interpreting tables. In Table 5.1, a weak candidate might notice only the difference in composition between inspired air and expired air. The table also shows the difference between alveolar air and expired air.

Once you have noted differences in the information in a table, make sure you can explain them.

Ventilation of the alveoli

Diffusion of gases between the atmosphere and the alveoli would be too slow to keep us alive. A ventilation process, which we commonly call breathing, increases the mass flow of these gases. Breathing involves:

- inspiration – air is pushed into the lungs from the atmosphere, providing fresh air to the alveoli
- expiration – air is pushed from the lungs to the atmosphere, removing stale air from the alveoli.

Table 5.2 summarises how inspiration and expiration are brought about by the action of muscles around the thoracic cavity.

- Diaphragm muscle within the diaphragm. When this contracts, it pulls the diaphragm downwards so that it compresses the contents of the abdomen below.
- External intercostal muscles attached to the outside of the ribs. When they contract, they pull the ribs upwards and outwards, increasing the volume of the thorax.

7 Suggest what would happen if the internal intercostal muscle, attached to the inside of the ribs, were to contract.

Stage of breathing	Diaphragm muscles	Intercostal muscles	Volume of thorax	Pressure in lungs
Inspiration – an active process	Contract	Contract	Increases as: ◆ diaphragm is pulled down ◆ ribs are pulled upwards and outwards.	Becomes less than that of atmosphere therefore air is pushed from atmosphere into lungs.
Expiration – a passive process	Relax	Relax	Decreases as: ◆ diaphragm is pushed upwards by elastic recoil of abdominal contents ◆ elastic tissue in lungs recoils ◆ gravity pushes ribs downwards.	Becomes greater than that of atmosphere therefore air is pushed from lungs into atmosphere.

TABLE 5.2 Inspiration and expiration are caused by pressure changes within the thoracic cavity resulting from the action of muscles attached to the diaphragm and the ribs

EXAMINER'S TIP

Remember that:
◆ air movement during breathing is caused by air being pushed from a high pressure to a lower pressure.
◆ the pressure of air increases when its volume decreases and decreases when its volume increases.

These concepts are important when describing, or interpreting data about, breathing.

Control of breathing

Figure 5.5 represents measurements of the breathing of someone at rest. It shows:
◆ a rhythmic pattern of inspiration and expiration
◆ that about 500 cm³ of air is inhaled and exhaled with each breath – this volume is called the tidal volume
◆ that the rhythmic pattern of breathing uses only a small part of the volume of the lungs.

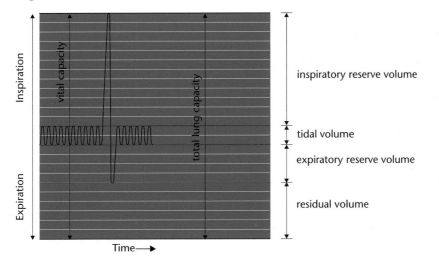

FIGURE 5.5 Breathing at rest has a rhythmical pattern involving only a small part of the volume of the lungs. Further air can be breathed in (**inspiratory reserve**) and out (**expiratory reserve**). Some of the air can never be breathed out (the **residual volume**).

Figure 5.6 summarises the control of this rhythmic pattern of breathing. Notice that the rhythm involves:

◆ a breathing centre in the medulla of the brain – this centre is divided into an inspiratory centre and an expiratory centre
◆ regular discharges of nerve impulses from the inspiratory centre to the diaphragm muscle along the phrenic nerve and to the intercostal muscles along the intercostal nerve
◆ temporary inhibition of the inspiratory centre by stretch receptors in the lungs.

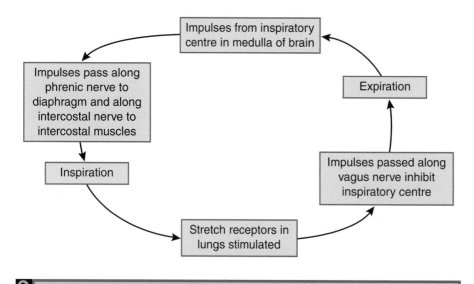

FIGURE 5.6 The rhythmical discharge of nerve impulses from the inspiratory centre of the medulla control breathing at rest.

8 In Figure 5.6, why does the inspiratory centre begin to send out impulses again after an expiration?

Ventilation rates

The amount of air that is breathed into and out of the lungs in one minute is called the minute ventilation rate. It can be calculated as:

Minute ventilation rate = Tidal volume × Breathing rate per minute

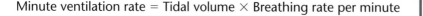

9 Calculate the minute ventilation rate of someone with a tidal volume of 500 cm³ and a breathing rate of 12 breaths per minute.

The minute ventilation rate is not constant. For example, when we exercise:
◆ the depth of each breath increases
◆ the rate of breathing increases.

These changes result from changes in the body associated with exercise, including:
◆ decreased pH (resulting from increased blood concentrations of CO_2), detected mainly by chemoreceptors in the respiratory centre of the medulla
◆ decreased blood concentrations of O_2, detected mainly by chemoreceptors in the carotid bodies and the aortic bodies (see Figure 5.7).

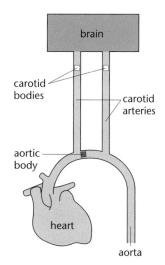

FIGURE 5.7 The positions of the aortic and carotid bodies.

? 10 Explain why:
a) exercise would increase the concentration of CO$_2$ in the blood,
b) an increased concentration of CO$_2$ in the blood would lower the blood's pH.

Figure 5.8 summarises the way in which changes in the carbon dioxide concentration of the blood affect breathing. You should recognise this as an example of homeostasis.

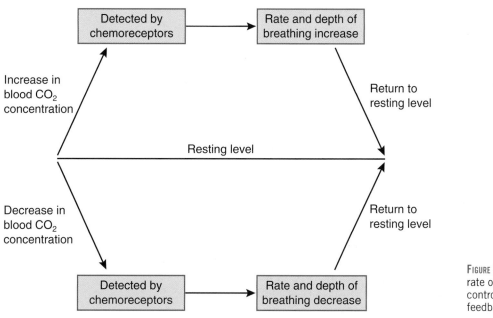

FIGURE 5.8 The depth and rate of breathing are controlled by negative feedback.

WORKED EXAM QUESTION

1   a)   An athlete exercised at different rates on an exercise bicycle. The table shows the effects of exercise rate on his breathing rate and tidal volume.

Exercise rate/ arbitrary units	Breathing rate/ breaths minute^{-1}	Tidal volume/ dm^3
0	14.0	0.74
30	15.1	1.43
60	15.3	1.86
90	14.5	2.34
120	15.1	2.76
150	14.8	3.25
180	21.5	3.21
210	25.7	3.23

(i)   The athlete cycled at the particular exercise rate for 5 minutes before the relevant readings were taken. Explain why the readings were taken only after the athlete had been cycling for 5 minutes.   *(1 mark)*

The readings were taken after 5 minutes to ensure that this was a fair test.

> This answer shows no understanding – answers including 'fair test' seldom do. A better answer would be that the athlete's breathing stabilises after 5 minutes.

(ii)   Calculate the total volume of air taken into the lungs in one minute at an exercise rate of 120 arbitrary units.   *(1mark)*

Volume = rate × tidal volume = 15.2 × 2.76 = 41.7 dm^3

> The candidate has clearly shown her working, chosen the relevant data from the table and correctly calculated the volume.

(iii)   Give **two** conclusions that can be drawn from the figures in the table.   *(2 marks)*

1   The breathing rate increases the harder the exercise.
2   The tidal volume increases as the exercise gets harder.

> The candidate has made a common mistake – she has not been precise enough in her answers. To gain marks, she needed to describe the two-fold nature of the patterns shown by the data. The breathing rate remains fairly constant until 180 arbitrary units, then increases. The tidal volume increases up to 150 arbitrary units then levels out. Make sure you look for, and describe, these patterns in data.

(AQA 2003)

EXAMINATION QUESTION

1   The graph shows the pattern of breathing in a person sitting at rest.

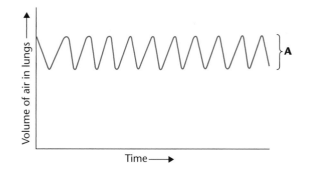

a) (i) What is the name given to the volume of air labelled **A**?   *(1 mark)*

   (ii) Explain how you would calculate the volume of air taken into the lungs in
        one minute.   *(1 mark)*

b) During exercise the breathing rate and the cardiac output both increase.

   (i) Describe how the medulla increases breathing rate.

   (ii) Describe how sympathetic nerves increase cardiac output.   *(4 marks)*

One way in which hospitals test how well the lungs are working is to measure the
gas transfer factor. This is done by measuring the uptake of carbon monoxide from
a single breath of air containing 0.3% carbon monoxide.

a) (i) By what process would carbon monoxide pass from the air in the alveoli to
        the blood in the lung capillaries?   *(1 mark)*

   (ii) Suggest why carbon monoxide is used for this test.   *(2 marks)*

d) Interstitial lung disease is a disease in which the alveolar walls become thicker.
    Explain why the gas transfer factor would be low in a person who had
    interstitial lung disease.   *(1 mark)*

(AQA 2002)

6

The heart and circulation

After revising this topic, you should be able to:

▶ relate the structure of blood vessels to their functions

▶ explain the role of capillaries in the formation of tissue fluid

▶ identify the main features of a mammalian heart

▶ describe the roles of the SAN, AVN and bundle of His in coordinating the heart beat

▶ describe the cardiac cycle and relate the main events to changes in volume and pressure

▶ explain how valves prevent backflow of blood

▶ explain the effect of exercise on changing heart rate and distribution of blood.

Blood vessels

A mammal's heart is a pump. From the heart, blood is circulated around the body in the various blood vessels.

◆ Blood leaves the heart in arteries. These branch into smaller arterioles and then into capillaries.

◆ In the capillaries, substances are exchanged between the blood and the cells of the body.

◆ Blood is collected from the capillaries by a network of venules. It flows from these venules into larger veins and back to the heart.

> **? 1** **In which one of these vessels is the pressure the greatest? Explain why.**

CIRCULATION

Figure 6.1 summarises the circulatory system of the human body.

> **EXAMINER'S TIP**
>
> You must remember the names of the main blood vessels as shown in Figure 6.1. You must also remember which organs these vessels go to and come from.

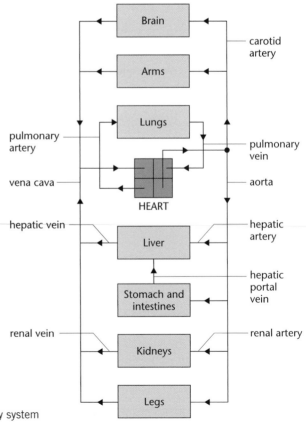

FIGURE 6.1 The circulatory system

STRUCTURE OF BLOOD VESSELS
Arteries, arterioles, venules and veins

Arteries, arterioles, venules and veins have the same basic structure. We only need to know the structure and function of arteries, arterioles and veins. Their walls have three layers and each layer has a specific function as shown in Table 6.1.

Feature	Artery	Arteriole	Vein
Appearance	outer layer, middle layer, inner layer ◆ Thick wall ◆ Narrow lumen	outer layer, middle layer, inner layer ◆ Thin wall	outer layer, middle layer, inner layer ◆ Thin wall ◆ Large lumen
Main function	◆ Transport blood from heart to organs	◆ Distribution of blood ◆ Transport blood in organs	◆ Transport blood to heart from organs
Blood flow	Away from heart to an organ	Within an organ to capillaries	Towards the heart away from an organ
Blood pressure	◆ High ◆ Pulsatile (pressure surges)	◆ High, but not as high as artery and less pulsatile	◆ Low ◆ Non-pulsatile
Valves	Absent	Absent	Present and prevent backflow of blood
Inner layer: endothelium	Present	Present	Present
Function	Very smooth layer which enables blood to flow with little friction		
Middle layer: muscle and elastic fibres	Present and thick	Present	Present but thinner than in artery
Function	When blood is forced into the arteries, they bulge outwards, stretching the elastic fibres. Recoil of these elastic fibres helps smooth out flow of blood. Contraction of the muscles in the walls of the arterioles allows the amount of blood flowing to the organs of the body to be varied.		
Outer layer: connective tissue	Present	Present	Present and thick
Function	Helps hold vessels open and prevent tearing during body movements		

TABLE 6.1 A summary of the structure and function of arteries, arterioles and veins

Capillaries

Capillaries have a different structure from the other blood vessels as shown in Table 6.2.

Feature	Capillary
Appearance	wall one cell thick ◆ Microscopic vessel ◆ Wall only one cell thick
Blood flow	Around cells of an organ
Blood pressure	Pressure drops throughout capillary network
Endothelium	Present
Function	As the endothelium is the only layer, the capillary is permeable to water and small molecules which move out of the capillary. This forms tissue fluid (see below) which allows the exchange of substances between blood and cells.

TABLE 6.2 A summary of the structure and function of capillaries

? **2** How does the capillary differ in structure from all of the other blood vessels?

EXAMINER'S TIP

Remember, it is the elastic tissue that stretches and recoils to smooth out blood flow, it is *not* muscle action.

Tissue fluid formation

Tissue fluid is the fluid that surrounds every cell of the body. It is formed from the plasma in the capillaries. Cells will obtain all their nutrients from the tissue fluid and will put into it all their waste products. The formation of tissue fluid is decribed in Table 6.3.

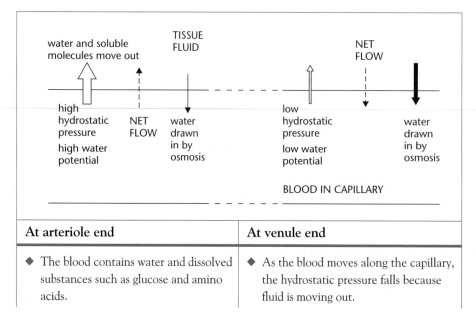

At arteriole end	At venule end
◆ The blood contains water and dissolved substances such as glucose and amino acids.	◆ As the blood moves along the capillary, the hydrostatic pressure falls because fluid is moving out.

◆ It also contains large suspended molecules such as plasma proteins, which are too large to leave the capillary.

◆ These plasma proteins *lower* the water potential.

◆ Water is drawn *into* the capillary by osmosis.

◆ However, blood flowing in the capillary has a high hydrostatic pressure caused by the contraction of the heart.

◆ The hydrostatic pressure forces water and small soluble molecules *out* through the walls of the capillary.

◆ The blood still contains plasma proteins.

◆ But the concentration is higher because water has been lost.

◆ Thus the water potential is even lower than that of the tissue fluid outside the capillary.

◆ Even more water is still drawn *into* the capillary by osmosis.

Result

The effect of the hydrostatic pressure is greater than that of the water potential so fluid is forced *out*, taking with it soluble nutrients from the blood.
This fluid now forms tissue fluid, which will bathe the cells.

Result

The effect of the water potential is greater than that of the hydrostatic pressure and water tends to be drawn back *into* the capillary, taking with it waste products produced by the cells.

TABLE 6.3 The formation and function of tissue fluid

EXAMINER'S TIP

◆ Fluid is moving into and out of the capillary all the time. Remember, more is *pushed out* at the arteriole end due to very high hydrostatic pressure and more is *drawn back in* at the venule end due to the large water potential gradient.

◆ Plasma is the name given to the fluid part of the blood. Remember, it is not plasma that moves out of the capillaries – it is water and small soluble molecules.

◆ Tissue fluid is the name given to the fluid that bathes the cells. Remember, it is not tissue fluid that is drawn back into the capillaries – it is water and small soluble waste products.

3 Describe the effect at the venous end of the capillary if a person has:
a) a high blood pressure
b) a very low protein diet.

Lymph

Lymph is the name given to the fluid transported by lymphatic vessels.

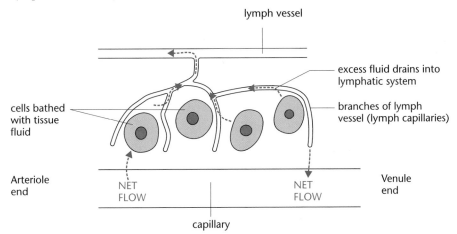

FIGURE 6.2 Movement of fluid in the lymphatic system

◆ Not all the fluid is returned to the blood at the venule end of the capillary.
◆ Some of the accumulated tissue fluid is drained into lymph vessels.
◆ When in the lymphatic system, the fluid is called lymph.
◆ The lymph vessels empty the lymph into the blood in the veins in the neck.

> **?** **4** **What are the differences in the composition of plasma, tissue fluid and lymph?**

Blood and blood cells

TISSUES AND ORGANS

Living organisms are made from cells.

◆ In organisms, similar cells are grouped together to form tissues.
◆ An organ consists of a number of different tissues. There are many obvious examples, such as the heart and the liver, and less obvious examples such as arteries and veins.

> **?** **5** **Name two tissues found in the vein.**

Blood is a tissue composed of plasma and cells. The cells are red blood cells and white blood cells, of which there is more than one type.

Plasma

◆ Plasma is the liquid part of blood.
◆ It contains soluble substances that are transported around the body, for example glucose, amino acids, hormones and minerals.
◆ As it is supplying and receiving substances from cells, the amount of substances present may vary.

Red blood cells

◆ The red blood cells transport respiratory gases, for example oxygen.
◆ The cells contain a pigment called haemoglobin. Haemoglobin combines with oxygen to form an unstable compound called oxyhaemoglobin. Because it is unstable, the haemoglobin will readily give up the oxygen to the cells.
◆ Red blood cells have no organelles such as a nucleus, mitochondria or ribosomes. This allows more room for haemoglobin.
◆ The cells are shaped like a biconcave disc. This means that they have a large surface area to volume ratio for the efficient diffusion of oxygen.

White blood cells

◆ The main function of the white blood cells is defence.
◆ They are larger than red blood cells and have all the organelles found in a eukaryotic cell.
◆ There are fewer white blood cells in the blood than red blood cells.
◆ Some types can leave the blood by squeezing through gaps in the capillary wall.

There are several types of white blood cell, as shown in Table 6.4.

Type of white blood cell	Appearance	Function
Lymphocyte	large, round nucleus — small amount of cytoplasm	◆ Secretes antibodies ◆ Involved in the humoral immune response
Monocyte	large, kidney-shaped nucleus	◆ Engulfs bacteria
Granulocyte	lobed nucleus — granular cytoplasm	◆ Engulfs bacteria (phagocytosis) ◆ Involved in the allergic response

TABLE 6.4 The different types of white blood cell

The heart

The heart is a pump which produces most of the pressure that moves blood through vessels in the body. It is made from **cardiac muscle**. This muscle does not fatigue which means it can contract and relax without tiring.

EXAMINER'S TIP

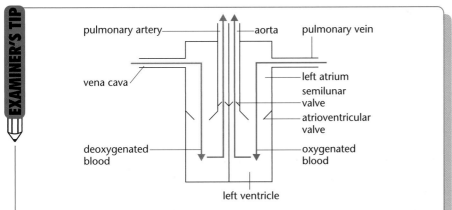

- ◆ There are many diagrams of the heart – make sure that the one you learn is simple but accurate like the one above. Though you will never be asked to draw a diagram of the heart in an exam, a simple diagram that can be drawn quickly is a good visual aide memoire to help you answer other questions.
- ◆ Learn not only the names of the structures in the heart but also the route that blood will take to the lungs and to the body.
- ◆ All arteries carry blood *away* from the heart, so the pulmonary artery goes from the right ventricle to the lungs. It is the only artery in the body that carries deoxygenated blood.

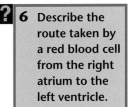

6 Describe the route taken by a red blood cell from the right atrium to the left ventricle.

CARDIAC CYCLE

The cardiac cycle is the sequence of events that lead to the filling and emptying of the heart.

Remember:

- ◆ muscle contraction *decreases* the volume in the heart chambers which will *increase* the pressure
- ◆ valves open and close maintaining unidirectional flow
- ◆ the right and left side of the heart fill and empty at the same time.

Although the cardiac cycle is continuous, it can be divided into the four main parts shown in Table 6.5.

Part of cardiac cycle	Description
Atrial diastole 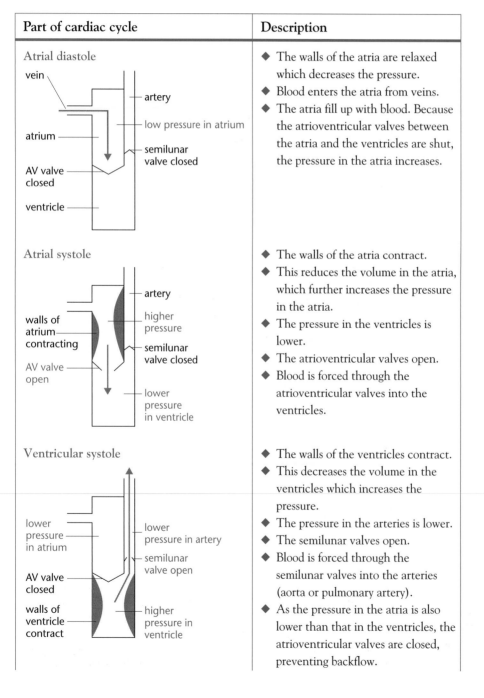	◆ The walls of the atria are relaxed which decreases the pressure. ◆ Blood enters the atria from veins. ◆ The atria fill up with blood. Because the atrioventricular valves between the atria and the ventricles are shut, the pressure in the atria increases.
Atrial systole	◆ The walls of the atria contract. ◆ This reduces the volume in the atria, which further increases the pressure in the atria. ◆ The pressure in the ventricles is lower. ◆ The atrioventricular valves open. ◆ Blood is forced through the atrioventricular valves into the ventricles.
Ventricular systole	◆ The walls of the ventricles contract. ◆ This decreases the volume in the ventricles which increases the pressure. ◆ The pressure in the arteries is lower. ◆ The semilunar valves open. ◆ Blood is forced through the semilunar valves into the arteries (aorta or pulmonary artery). ◆ As the pressure in the atria is also lower than that in the ventricles, the atrioventricular valves are closed, preventing backflow.

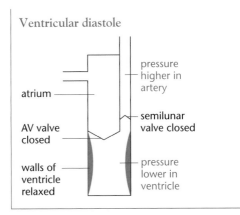

Ventricular diastole

atrium

AV valve closed

walls of ventricle relaxed

pressure higher in artery

semilunar valve closed

pressure lower in ventricle

◆ The walls of the ventricles relax, which decreases the pressure in the ventricles.
◆ As the pressure in the ventricles is lower than the pressure in the arteries, the semilunar valves are closed, preventing backflow.

TABLE 6.5 The four main parts of the cardiac cycle

WHAT CAUSES THE HEART VALVES TO OPEN AND CLOSE?

As the heart muscle contracts, the pressure in the chambers increases. The heart valves ensure that blood moves only in one direction, from the atria to the ventricles to the arteries. Table 6.6 summarises the action of the cardiac valves.

Atrioventricular valves open because	Atrioventricular valves open because	Semilunar valves open because	Semilunar valves close because
pressure is higher in the atria than the ventricles. Blood will flow from atria to ventricles.	pressure is higher in the ventricles than the atria. They prevent blood flowing back into the atria when the ventricles contract.	pressure is higher in the ventricles than the arteries. Blood will flow from the ventricles into the arteries.	pressure is higher in the arteries than in the ventricles. They prevent backflow of blood into the ventricles.

TABLE 6.6 The action of the cardiac valves

PRESSURE GRAPHS

The cardiac cycle is often represented as a graph as shown in Figure 6.3.

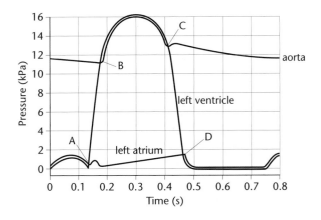

FIGURE 6.3 A pressure graph of the cardiac cycle

EXAMINER'S TIP

◆ It is important that when you are interpreting a pressure graph you realise that when the lines of the graph cross, this indicates that pressure in one area is greater than in another. This causes a valve to close or open, shown as points A, B, C and D in Figure 6.3.
◆ Calculation of heart rate from a pressure graph is often asked. Remember to work out the time from an obvious point, such as when a valve closes, to when it happens again. Do not go back to zero.
◆ Make sure you read off any figures from the graph accurately.
◆ The information from a pressure graph can also be given in the form of a table or as volume changes. Make sure you are familiar with all these examples.

7 The heart sounds are described as a 'lub dup'.
   a) What causes the heart sounds?
   b) Using Figure 6.3, at what time from the beginning of the cardiac cycle do the heart sounds occur?
   c) Calculate the heart rate from the graph.

COORDINATING HEART BEAT

The heart muscle does not have to be stimulated by a nerve before it will contract. The heart beat originates in the muscle itself. We say that heart muscle is myogenic.

SAN	◆ The heart beat starts with an impulse being produced at the sinoatrial node, SAN.
SAN	◆ The impulse spreads through the muscle of the atria. ◆ The impulse causes the atria to contract. ◆ Blood is forced from the atria into the ventricles.

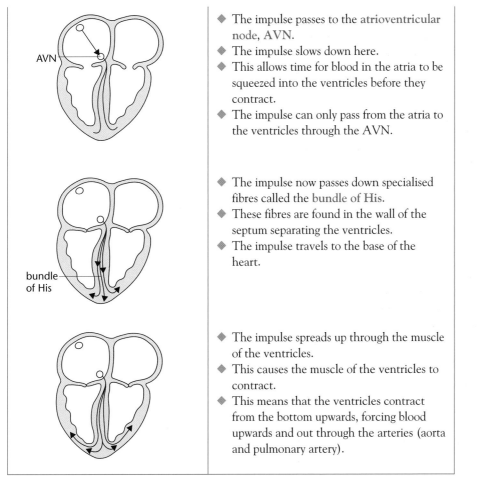

The impulse passes to the atrioventricular node, AVN.
- The impulse slows down here.
- This allows time for blood in the atria to be squeezed into the ventricles before they contract.
- The impulse can only pass from the atria to the ventricles through the AVN.

- The impulse now passes down specialised fibres called the bundle of His.
- These fibres are found in the wall of the septum separating the ventricles.
- The impulse travels to the base of the heart.

- The impulse spreads up through the muscle of the ventricles.
- This causes the muscle of the ventricles to contract.
- This means that the ventricles contract from the bottom upwards, forcing blood upwards and out through the arteries (aorta and pulmonary artery).

TABLE 6.7 How the heart beat is coordinated

EXAMINER'S TIP

It is perfectly acceptable to use SAN and AVN in your exam as they are standard abbreviations.

THE EFFECT OF EXERCISE

- Stroke volume: this is the amount of blood that the left ventricle pumps out each time it beats.
- Heart rate: this is the number of times the heart beats in one minute.
- Cardiac output: this is the product of the stroke volume and the heart rate. It is often represented as a simple equation:

$$\text{Cardiac output} = \text{Stroke volume} \times \text{Heart rate}$$

When undertaking exercise it is important that the cells of the body receive sufficient oxygen, and that nutrients and waste products such as carbon dioxide are taken away.

There are three ways in which the cardiac output can be changed. These are described in Table 6.8.

Hormones	◆ Adrenaline is secreted. ◆ This travels in the blood. ◆ It increases the rate at which the SAN sends impulses. ◆ It increases heart rate.

Nerve impulse from the brain
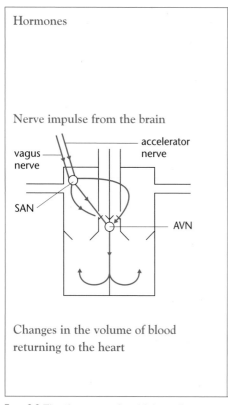
	◆ Receptors monitor carbon dioxide levels in the blood.
	◆ Nerves from these receptors carry impulses to the cardiovascular centre in the brain.
	◆ There are two nerves from this centre:
	  – accelerator nerve: impulses travelling down this nerve to the SAN increase the heart rate
	  – vagus nerve: impulses travelling down this nerve to the SAN slow down the heart rate.

Changes in the volume of blood returning to the heart	If more blood enters the heart and stretches its walls more than normal, the heart responds by beating faster and with greater strength.

TABLE 6.8 The three ways in which cardiac output can be changed

Effects of exercise on circulation

- ◆ More exercise means a faster rate of respiration and the need for more oxygen and nutrients to be supplied to the muscles.
- ◆ This can be achieved by increasing the amount of blood flowing through the capillaries that supply muscles.
- ◆ A large increase in blood flowing to one part of the body must be met by a reduction in the amount of blood supplied to other parts of the body.
- ◆ During exercise there is an increase in the amount of blood flowing to the:
  - – muscles – allowing increased respiration
  - – skin – allowing increased removal of excess heat.
- ◆ There is a decrease of blood flowing to the organs that make up the digestive system.
- ◆ The brain needs a constant supply of oxygen so its blood supply is never affected.

WORKED EXAM QUESTION

1   The diagram shows a section through a
    human heart.

    a) Which of the blood vessels labelled **A** to **D**

        (i) takes blood from the heart to the
            muscles of the arms and legs;

                                        (1 mark)

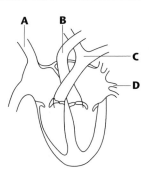

        The aorta

> The examiners have made it easy to get this mark, even if the candidate could not
> remember the name of the vessel. The answer expected was vessel B. He has instead
> named the correct artery and will be awarded the mark.

        (ii) is a vein which contains oxygenated blood?   *(1 mark)*

            The only vein that contains oxygenated blood comes back to
            the heart from the lung. It is D (the hepatic vein).

> Although the candidate obviously knows the right vein, i.e. vessel D which would
> have got him the mark, he has attempted once again to name it and has made a
> mistake. He should have named the pulmonary vein. Never give any extra
> information unless you are sure that it is correct. If you give one right and one wrong
> answer the examiner will not make the choice and you will not gain the mark.

    b) Is the right ventricle filling with blood or emptying? Give **two** pieces of
        evidence from the diagram to support your answer.   *(2 marks)*

        Filling or emptying?   *Filling*

        Evidence

        1. *The bicuspid valve is open*

        2. *The semilunar valve is closed*

> This candidate has remembered the direction of blood flow and has realised that as the
> valve into the ventricle is open, blood must be entering the chamber. The closed valve
> from the ventricle to the artery means that the blood cannot leave and thus the
> chamber is filling. However, he has named the valve between the left atrium and
> ventricle instead of the one between the right atrium and ventricle, which is the
> tricuspid valve. Bicuspid and tricuspid are acceptable names for the valves, but it is
> easier to call the valves between the atria and ventricles the atrioventricular valves,
> which applies to both sides of the heart. Remember to read every word of the question.

The graph shows some of the changes in pressure and volume during part of a cardiac cycle.

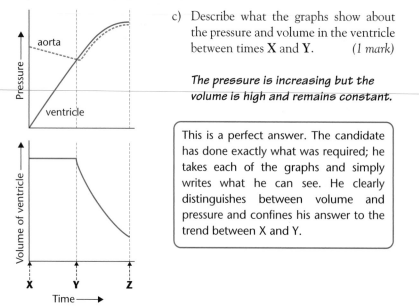

c) Describe what the graphs show about the pressure and volume in the ventricle between times **X** and **Y**. *(1 mark)*

The pressure is increasing but the volume is high and remains constant.

This is a perfect answer. The candidate has done exactly what was required; he takes each of the graphs and simply writes what he can see. He clearly distinguishes between volume and pressure and confines his answer to the trend between X and Y.

d) At point Y, the valve between the aorta and the ventricle opens. Use the information about pressure on the graph to explain why. *(1 mark)*

The pressure in the ventricle is now high enough to open the valve.

The actual pressure is not important here and so this answer is wrong. The valve opens because the pressure is higher in the ventricle than in the aorta and that is what the graph shows at point Y.

e) Describe the structure of a capillary and explain how capillaries are important in the formation of tissue fluid. *(6 marks)*
- *The wall of the capillary is one cell thick*
- *It is made of squamous epithelium*
- *Which allows plasma to pass through*
- *At the arterial end the hydrostatic pressure is high*
- *Plasma is forced out*
- *Forming tissue fluid*

Using tables, diagrams and bullet points is perfectly acceptable at AS. This candidate has selected six ideas and presumes that each will be worth a mark. However, plasma does not pass through the wall of the capillary, water and small soluble molecules do, so this mistake has lost two potential marks. The final point only restates the question and is also not worth a mark. The answer lacks enough detail to gain more than three marks. He has forgotten the effect of plasma proteins in lowering the water potential of the blood. As they are large molecules they do not pass through the wall of the capillary, and thus water would be removed by osmosis from the existing tissue fluid. The fact that the high hydrostatic pressure has a greater effect than osmosis causing a net movement out of the capillary has not been mentioned.

(AQA 2003)

EXAMINATION QUESTION

1   The diagram shows the heart and the tissues which control the heart beat. The figures on the diagram show the time in seconds taken for a wave of electrical activity to spread from the sinoatrial node.

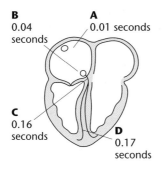

a) (i) The diagram shows that the valve between the right atrium and the right ventricle is closed. What does this indicate about the relative pressures in the right atrium and the right ventricle? *(1 mark)*

    (ii) Copy and complete the diagram to show whether the valve between the left atrium and the left ventricle is open or closed. *(1 mark)*

b) The rate at which the electrical activity passes from **B** to **C** is important in controlling the heart beat. Explain why. *(2 marks)*

c) The heart rate of a sleeping person is low. Explain how nerves supplying the heart may produce a low heart rate in a sleeping person. *(3 marks)*

(AQA 2003)

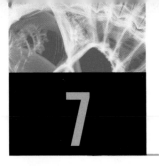

7 Enzyme technology

After revising this topic, you should be able to:

▶ distinguish between intracellular enzymes and extracellular enzymes

▶ explain how microorganisms can be grown and the enzymes they produce can be extracted and purified on a commercial scale

▶ demonstrate an understanding of the properties of enzymes that make them suitable as analytical reagents

▶ demonstrate an understanding of the value of immobilised enzymes in industrial processes.

This topic builds on your understanding of enzymes. You should revise the content of Chapter 4 before working through this chapter.

The value of using isolated enzymes in industry

Many of the chemical reactions used in industry can be speeded up in one of two ways:

1 using inorganic catalysts, high temperatures or high pressures
2 using a specific enzyme.

Table 7.1 summarises the advantages and disadvantages of using enzymes.

Advantages of using enzymes	Explanation
◆ Enzymes are specific.	◆ Since enzymes react with specific substrates to produce specific products, there are less likely to be unwanted by-products formed.
◆ Most enzymes work at atmospheric pressure, moderate temperatures and near-neutral pH.	◆ Less money need be spent on energy to raise temperatures or pressures.
◆ Enzymes are proteins.	◆ Proteins are biodegradable, so cause little pollution.

Disadvantages of using enzymes	Explanation
◆ Enzymes are usually present in a mixture of other compounds.	◆ Purifying enzymes is an expensive process.
◆ Enzymes are sensitive to temperature and pH, making them unstable.	◆ Slight changes in conditions might denature the enzymes being used. More of the purified enzyme must be produced to replace that which has been lost.
◆ Enzymes are sensitive to inhibitors.	◆ Conditions must be free of contamination.

TABLE 7.1 The advantages and disadvantages of using enzymes in industrial processes

? **1** What makes enzymes specific to their substrate?

Production of industrial enzymes

Enzymes for industrial use are commercially produced from microorganisms. Figure 7.1 summarises the major stages in the commercial production of enzymes. You need to be familiar with only two of these: fermentation and downstream processing.

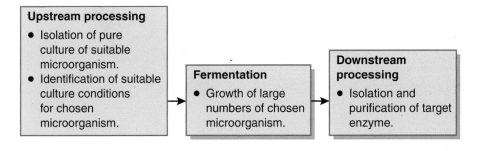

Upstream processing
- Isolation of pure culture of suitable microorganism.
- Identification of suitable culture conditions for chosen microorganism.

Fermentation
- Growth of large numbers of chosen microorganism.

Downstream processing
- Isolation and purification of target enzyme.

FIGURE 7.1 The main stages in the production of industrial enzymes. You need be familiar only with fermentation and downstream processing.

Growth of large numbers of microorganisms

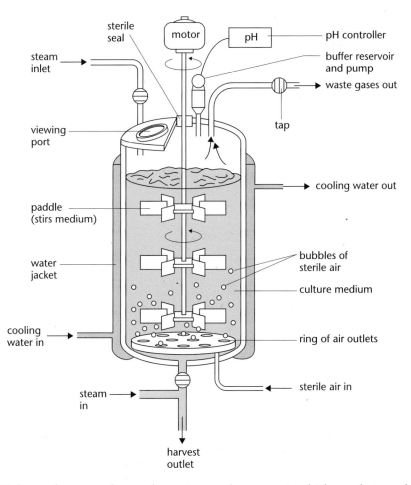

FIGURE 7.2 A simplified diagram of a closed industrial fermenter. During batch fermentation, a fermenter such as this is used to culture microorganisms in a fixed volume of medium for a fixed length of time before it is emptied.

Figure 7.2 shows a large, stainless steel container, or fermenter, in which populations of enzyme-producing microorganisms are grown. Table 7.2 summarises the function of different parts of the fermenter.

Part of fermenter	Further notes
Outer water jacket	Contains flowing cold water and cools contents of fermenter; heat is released by respiration of microorganism.
Stainless steel cylinder	Forms the body of the fermenter. Stainless steel is stronger than mild steel and so is self-supporting and able to withstand the high internal pressures when gases are produced during fermentation. Its surfaces are easily polished to a smooth finish that will not form pockets in which microorganisms can collect. After being cleaned and sterilised, the fermenter is filled with a sterile nutrient solution, which is inoculated with a pure culture of the bacterium or fungus to be grown.
Motor	Drives the paddles which mix the nutrient and the microorganisms.
Air inlet	Allows entry of sterile air when fermentation is aerobic.
Pressure release valve	Positioned above the liquid level, allows gases released by microbial respiration to escape through a gas exhaust system.
Foam breaker	Removes foam that can form during fermentation. Foam often causes loss of culture into the gas exhaust system and can clog the gas exhaust filter.
Nutrient inlet	Allows fresh sterile medium to be added to replace the nutrients that have been used up.
Temperature and pH monitors	Allow temperature and pH inside fermenter to be checked and adjusted.
Sample tube	Allows small samples to be removed for analysis.
Tap	Allows emptying of fermenter.

TABLE 7.2 The components of an industrial fermenter for enzyme production. After inoculation, the microbial population is allowed to grow. At the required time, the fermenter is emptied and the desired products isolated.

2 Explain why a water jacket is needed around a fermenter.

EXAMINER'S TIP

If asked for an explanation, ensure you give one. To answer 'to cool the fermenter' in response to Question 2 would not be accepted as an explanation. You need to explain why the temperature might increase – where is the heat likely to come from – or explain the detrimental effect of a rise in temperature.

Culture media

Like humans, microorganisms have specific nutritional requirements. These are not the same for different types of microorganism, so culture media are specific to each microorganism used. The culture medium must contain the following components in an appropriate concentration:

- an energy source, such as carbohydrate
- a source of carbon – for some this can be carbon dioxide, for others it must be an organic compound such as carbohydrate
- a source of nitrogen
- inorganic ions
- oxygen, if the microorganism respires aerobically.

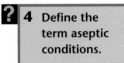

3 Explain why microorganisms need a supply of:
a) carbon
b) nitrogen.

> **EXAMINER'S TIP**
>
> The specification states that a suitable example should be chosen to show the growth of large numbers of microorganisms and the isolation and purification of an enzyme. However, given the vast number of enzymes produced commercially, it would be almost impossible to devise a mark scheme that would cover them all. It is likely that examination questions will *either* test your understanding of general principles *or* give you data and expect you to interpret them using your understanding of general principles.

Aseptic conditions

For successful fermentation, it is vital that no unwanted microorganisms enter the fermenter. It is also vital that microorganisms do not 'escape' from the fermenter. Aseptic techniques ensure that both these requirements are met. Table 7.3 summarises the main ways in which aseptic conditions are maintained in a fermenter.

4 Define the term aseptic conditions.

Feature	Explanation of how this technique ensures aseptic conditions are maintained
Washing, disinfecting and steam cleaning the fermenter and its inlet and outlet pipes	Washing removes surplus culture medium. Disinfecting kills microorganisms with suitable chemicals. The high temperature of steam kills microorganisms.
All inlet and outlet pipes have filters	Filters trap microorganisms, preventing their entry into, or their escape from, the fermenter.
All joints are sealed	Prevents entry of microorganisms into, or their escape from, the fermenter through faulty joints.
Polishing surfaces of fermenter	Rough surfaces could act as reservoirs for microorganisms.
Culture medium sterilised	Ensures there are no live microorganisms in the medium before it is put into the fermenter.

TABLE 7.3 The major precautions that ensure aseptic conditions in an industrial fermenter. Without these precautions, undesirable microorganisms could enter the fermenter and grow in the medium, producing undesirable products, or microorganisms could escape from the fermenter and contaminate other areas.

Isolation and purification of desired enzyme

The enzymes produced by microorganisms can be:

◆ intracellular, i.e. they are retained within the cells that produce them and speed up reactions which occur within these cells

◆ extracellular, i.e. they are secreted from the cells that produce them and speed up reactions which occur outside these cells.

Whether the desired enzyme is intracellular or extracellular affects its isolation and purification, as shown in Figure 7.3. The isolation and purification of the desired enzymes in this flow chart is called downstream processing.

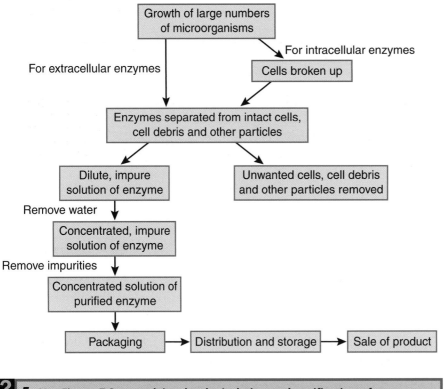

EXAMINER'S TIP

'Downstream processing' is a subject-specific term. It has nothing to do with the usual meaning of downstream that you might use in ecology.
Use Figure 7.1 to make sure you can explain what downstream processing is in an examination.

FIGURE 7.3 The isolation and purification of the enzyme is called downstream processing and is by far the most expensive stage in producing enzymes.

> **? 5** Use Figure 7.3 to explain why the isolation and purification of an intracellular enzyme is usually more expensive than that of an extracellular enzyme.

The use of enzymes as analytical reagents

Biosensors are used to test for the presence of particular compounds. For example, they are used to test the alcohol content of the blood of drink-driving suspects and to test the urine of athletes for the presence of banned drugs.

Enzymes are useful in biosensors because they are:

◆ highly sensitive – enabling the detection of very low concentrations of their substrate

◆ highly specific – enabling detection of their substrate and no other similar compound.

Figure 7.4 shows the tip of an electrode used in a biosensor that monitors urea in blood or urine. It contains urease, an enzyme that catalyses the following reaction:

$$\text{urea} + \text{water} \xrightarrow{\text{urease}} \text{carbon dioxide} + \text{ammonium ions}$$

Figure 7.5 summarises how this biosensor works.

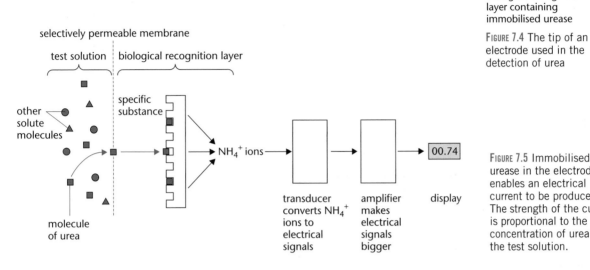

FIGURE 7.4 The tip of an electrode used in the detection of urea

FIGURE 7.5 Immobilised urease in the electrode enables an electrical current to be produced. The strength of the current is proportional to the concentration of urea in the test solution.

Immobilised enzymes

In your coursework, you mixed solutions of an enzyme with solutions of its substrate. This can also be done in commercial operations. This is, however, a wasteful use of enzymes since:

◆ the enzyme is sensitive to the effects of temperature and pH changes
◆ the product is contaminated by the enzyme, which is costly to remove
◆ the enzyme solution must be replaced each time new substrate is added.

These problems are overcome by using immobilised enzymes, i.e. pure enzymes immovably bound to a fixed surface. Figure 7.6 shows some of the ways that enzymes can be immobilised. Table 7.4 summarises the advantages of using immobilised enzymes over the use of solutions of enzymes.

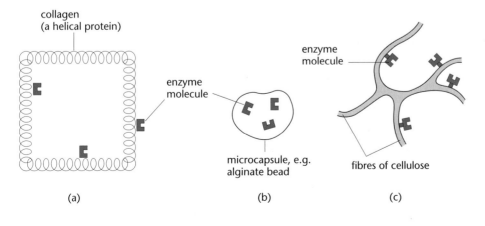

FIGURE 7.6 Pure enzymes can be immobilised in several ways. Enzymes can be:
(a) adsorbed on to a collagen matrix;
(b) trapped within microcapsules, such as alginate beads;
(c) attached to cellulose fibres.

Feature	Enzymes in solution	Immobilised enzymes
Contamination of product	Enzymes are mixed with product. Decontamination is time consuming and expensive.	Enzymes are held in a matrix and so do not mix with the product. Time and money spent on decontamination is saved.
Re-use of enzyme	Enzyme is lost with product or is expensive to recover and purify from enzyme–product mixture.	Since it is not mixed with the product, the enzyme can be easily and cheaply recovered and re-used.
Control of reaction	Enzyme cannot be separated from reactants, so reaction cannot easily be stopped.	If held on a matrix, enzyme can be removed from reactants by removing the matrix from the reactant mixture. This allows reaction to be stopped and started as the need arises.
Stability of enzyme in industrial processes that involve high temperatures or changes in pH	An enzyme operates in a narrow range around its optimum temperature and pH. The reaction it catalyses might release heat or change the pH of the solution. Increases in temperature or changes in pH would reduce enzyme activity.	The matrix around them protects immobilised enzymes. This makes immobilised enzymes more stable to high temperatures (more thermostable) or changes in pH.

TABLE 7.4 A comparison of the commercial use of free enzymes and of immobilised enzymes. Immobilisation is commonly used in industry because its advantages outweigh its disadvantages, particularly because immobilisation increases the thermostability of enzymes and removes the need to separate enzymes from the desired product.

? 6 Explain why enzymes used in industrial processes must have a high degree of thermostability.

WORKED EXAM QUESTION

1 The enzyme cholesterol oxidase is used to measure cholesterol concentration in blood samples. The fungus *Aspergillus niger* produces cholesterol oxidase. The flow chart summarises the procedures used to isolate and purify cholesterol oxidase.

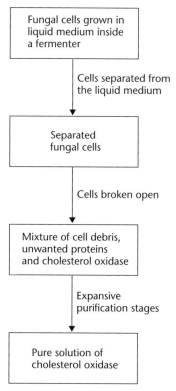

```
┌─────────────────────┐
│ Fungal cells grown in │
│ liquid medium inside  │
│ a fermenter           │
└─────────────────────┘
          │  Cells separated from
          │  the liquid medium
          ▼
┌─────────────────────┐
│ Separated           │
│ fungal cells        │
└─────────────────────┘
          │
          │  Cells broken open
          ▼
┌─────────────────────┐
│ Mixture of cell debris, │
│ unwanted proteins       │
│ and cholesterol oxidase │
└─────────────────────┘
          │
          │  Expansive
          │  purification stages
          ▼
┌─────────────────────┐
│ Pure solution of    │
│ cholesterol oxidase │
└─────────────────────┘
```

a) (i) What term describes the stages of isolating and purifying an enzyme produced in a fermenter? (*1 mark*)

This is the industrial use of enzymes.

> The candidate has failed to recall a specific term. The answer is 'downstream processing'.

(ii) Suggest one method which could be used to separate the fungal cells from the liquid medium. (*1 mark*)

This is called chromatography.

> Again, the candidate has failed to recall an appropriate term. A mark would have been awarded for any one of the following: centrifugation; ultracentrifugation; filtration.

b) Why must amino acids be present in the liquid medium inside the fermenter?

(*1 mark*)

The fungus needs these to make the enzyme, which is a protein.

As often occurs, this candidate has failed to recall specific terms in part a) but shows understanding of biological principles here. In fact, a mark would have been given for 'to make proteins' or 'to make the enzyme'.

c) What evidence from the flow chart suggests that cholesterol oxidase is an intracellular enzyme? (*1 mark*)

The cells have to be broken up, which suggest the enzyme is not in the medium already.

The candidate has selected the correct evidence from the flow chart and gains the mark.

d) The fungal cells used to produce the cholesterol oxidase are grown in aseptic conditions. This prevents the growth of unwanted microorganisms in the liquid medium. Suggest and explain **one** reason why the growth of unwanted microorganisms would make cholesterol oxidase more expensive to produce.

(*2 marks*)

The unwanted microorganisms would have to be destroyed.

The candidate has produced a weak suggestion and has not attempted an explanation. There are several ways of answering this question correctly. The unwanted microbes would compete for nutrients and lower the yield, the unwanted microorganisms might produce toxins that would destroy the fungus or the unwanted microorganisms would produce unwanted proteins that would have to be separated from the desired enzyme.

(*AQA 2001*)

EXAMINATION QUESTION

1   The diagram shows a fermenter used to grow large numbers of microorganisms for enzyme production. To achieve rapid growth of these microorganisms, particular conditions are needed.

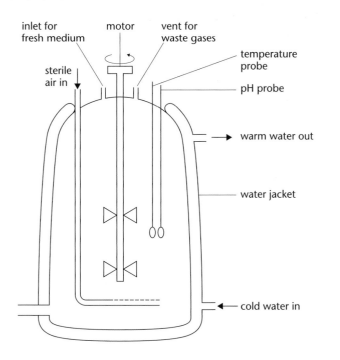

a) (i) Explain why the pH of the culture is kept constant.   *(2 marks)*

    (ii) Suggest why a water jacket is necessary.   *(2 marks)*

b) Aseptic conditions are used in the production of enzymes. Explain how contamination with other microorganisms might affect the amount of enzyme produced.   *(2 marks)*

c) Enzymes produced by microorganisms may be intracellular or extracellular. Explain why intracellular enzymes are usually more expensive to produce commercially.   *(2 marks)*

(AQA 2002)

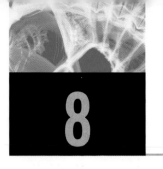

8 DNA and protein synthesis

After revising this topic, you should be able to:

▶ describe the structure of DNA and RNA

▶ understand the evidence that DNA is the genetic material

▶ understand how DNA replicates

▶ explain protein synthesis

▶ understand how DNA has an effect on the phenotype.

The basic structure of DNA

DNA (deoxyribonucleic acid) is a polymer, the monomer of which is a nucleotide. Each nucleotide is made of a nitrogenous base, a deoxyribose sugar and a phosphate. These can be represented by the symbols shown in Table 8.1.

Components of a nucleotide	Represented by
Nitrogenous base	
Deoxyribose sugar	
Phosphate	

TABLE 8.1 Simplified representations of the parts of a nucleotide

These three molecules are joined together by covalent bonds, resulting in a molecule which does not break down easily.

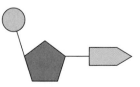

FIGURE 8.1 A simple representation of a nucleotide

JOINING NUCLEOTIDES TOGETHER

◆ Nucleotides join together to form a chain.
◆ There are four different nucleotides, each with a different nitrogenous base.
◆ The bases are adenine, guanine, cytosine and thymine.
◆ The order of the nucleotides varies in each chain of DNA.

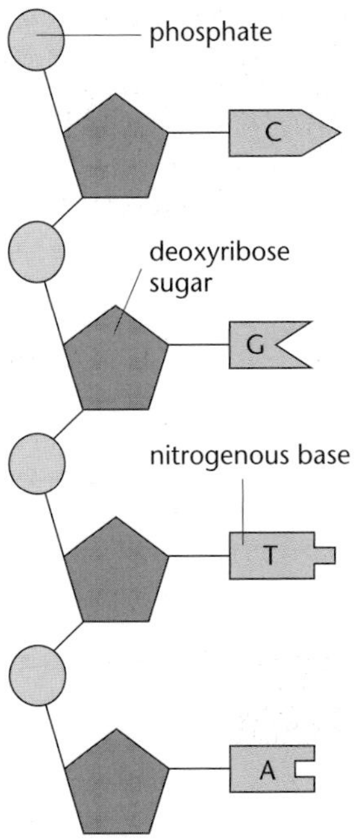

EXAMINER'S TIP

Remember, at AS you have to give details, so not just a base, a **nitrogenous base**, not just a sugar, a **deoxyribose sugar** and especially NOT phosphorus, a **phosphate**.
Be accurate when you refer to the components of a nucleotide.

? 1 **What other biological molecules are polymers, and what are the monomers from which they are made?**

DNA, A DOUBLE HELIX

◆ Pairs of nitrogenous bases are complementary – they join with each other.
◆ Two chains of nucleotides are linked by complementary nitrogenous bases.
◆ They join by 'weak' hydrogen bonds.
◆ Adenine pairs with thymine; cytosine pairs with guanine.
◆ The ratios of the complementary bases are always 1:1 in every DNA molecule.
◆ The ratios of any pair of non-complementary bases vary in every DNA molecule.

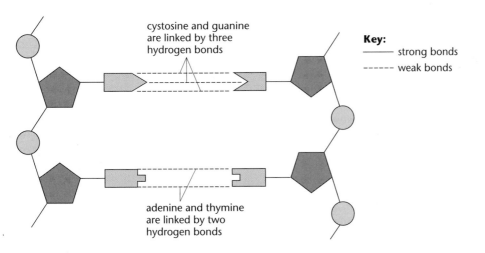

phosphate

deoxyribose sugar

nitrogenous base

FIGURE 8.2 A single chain of nucleotides. The letters represent the bases adenine (A), cytosine (C), guanine (G) and thymine (T).

cystosine and guanine are linked by three hydrogen bonds

Key:
—— strong bonds
----- weak bonds

adenine and thymine are linked by two hydrogen bonds

FIGURE 8.3 How the bases pair in a DNA double helix

EXAMINER'S TIP

◆ DNA codes for the production of protein, it does not contain any protein itself.
◆ Exam questions will represent the different molecules which make up nucleotides in different ways – be ready for this.

STRONG BONDS AND WEAK BONDS

◆ The strong bonds between the nucleotides in a chain maintain the order of the nucleotides – the code does not change.
◆ The weak bonds between the complementary bases allow the two chains to separate – during replication and transcription.

OTHER NUCLEIC ACIDS

DNA is not the only nucleic acid. Another form is called ribonucleic acid or RNA. There are similarities and a number of differences between DNA and RNA. These are summarised in Table 8.2.

Feature	DNA	RNA
Sugar in the nucleotide	Deoxyribose	Ribose
Bases in the nucleotide	Cytosine, guanine, adenine, thymine	Cytosine, guanine, adenine, uracil (thymine has been replaced by uracil)
General structure	Double strand	Single strand

TABLE 8.2 Comparing DNA and RNA

Table 8.3 gives some further differences between the two nucleic acids.

Feature	DNA	RNA
Stability	Very stable and long lasting	Unstable and breaks up quickly
Situation	Only found in the nucleus of a eukaryotic cell	Found in both the nucleus and the cytoplasm
Varieties	Only one form of DNA	Different varieties of RNA: mRNA, tRNA

TABLE 8.3 Further differences between DNA and RNA

ROLES OF THE NUCLEIC ACIDS

DNA

DNA is the code for the order of amino acids in a number of polypeptides.
(A gene is a small section of DNA that codes for a single polypeptide.)

mRNA

Messenger RNA (mRNA) is a complementary copy of the section of DNA which codes for one polypeptide. It takes the code from the nucleus to the site of protein synthesis, a ribosome.

tRNA

Transfer RNA (tRNA) carries a specific amino acid to the ribosome where the amino acids are joined by peptide bonds to form a polypeptide.

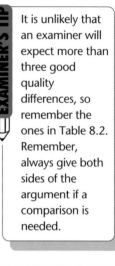

? 2 Bonds between nucleotides are formed by a condensation reaction. Explain the term condensation.

EXAMINER'S TIP It is unlikely that an examiner will expect more than three good quality differences, so remember the ones in Table 8.2. Remember, always give both sides of the argument if a comparison is needed.

? 3 What are the similarities between DNA and RNA?

4 Explain why DNA is stable whilst RNA is not stable.

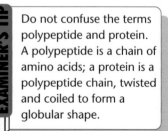

EXAMINER'S TIP Do not confuse the terms polypeptide and protein. A polypeptide is a chain of amino acids; a protein is a polypeptide chain, twisted and coiled to form a globular shape.

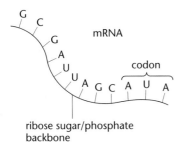

mRNA

codon

ribose sugar/phosphate
backbone

A straight, single-stranded chain

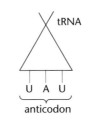

tRNA

U A U

anticodon

A folded, single-stranded chain

FIGURE 8.4 The structure of a molecule of mRNA and a molecule of tRNA

DNA and protein synthesis

TRANSCRIPTION

In the process of transcription, DNA is used as a template to produce a molecule of mRNA. This occurs in the nucleus. Transcription is summarised in Table 8.4.

EXAMINER'S TIP

'Transcribe' means to copy. A line of bases making up a part of a DNA molecule is copied into a line of bases making up an mRNA molecule.

Stage in transcription process	Diagram
◆ DNA is unwound. ◆ Hydrogen bonds between the two strands are broken.	hydrogen bonds are broken this section 'represents' the gene that codes for a polypeptide A═T C G G C A T T A C≡G sense strand non-sense strand
◆ The sense strand is used as a template. ◆ Complementary RNA nucleotides link with the exposed DNA nucleotides. ◆ The enzyme RNA polymerase joins together the RNA nucleotides.	RNA polymerase enzyme A═U C≡G G◄ C A A◄ U T C free RNA nucleotides in the nucleus sense strand bases of non-sense strand are not shown

◆ Hydrogen bonds attaching mRNA to the sense strand are broken.
◆ mRNA is released from the DNA.
◆ mRNA leaves the nucleus via the nuclear pores.

◆ In the cytoplasm mRNA enters the ribosome.

◆ Complementary bases of the DNA rejoin.

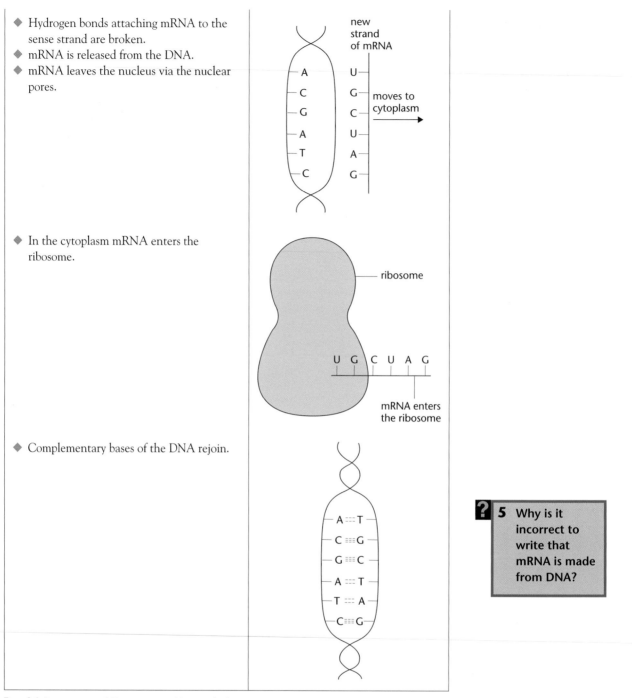

5 Why is it incorrect to write that mRNA is made from DNA?

TABLE 8.4 A summary of the process of transcription

TRANSLATION

In the process of **translation**, mRNA is used as a template to indicate the order of amino acids in a polypeptide. This is summarised in Table 8.5.

Stage in translation process	Diagram
A 'frame' of three bases of the mRNA molecule (this is called a **codon**) enters a ribosome.	
A complementary set of three bases on a tRNA molecule (this is called an **anticodon**) link to the codon by hydrogen bonds.	
Each type of tRNA is linked to a specific amino acid.	
◆ Codon/anticodon matching is repeated for a second 'frame' of codons. ◆ The two amino acids join by a peptide bond. ◆ The first amino acid detaches from the tRNA.	

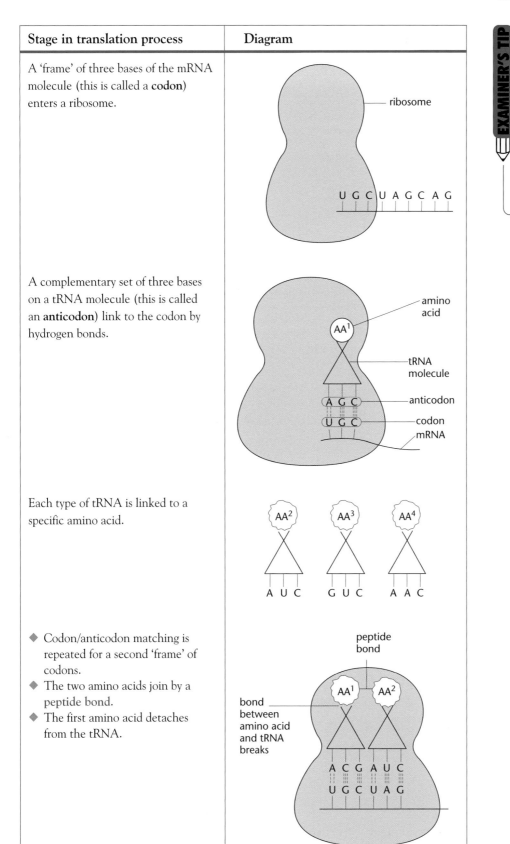

EXAMINER'S TIP

'Translate' means to change from one form to another.
A line of bases making up an mRNA molecule is changed into a line of amino acids making up a polypeptide.

- This process continues until a polypeptide is created.
- The polypeptide twists forming different bonds between the R groups of the amino acids to produce a protein.
- Freed tRNA returns to cytoplasm to join with another specific amino acid.

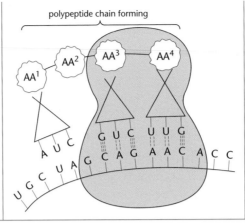

TABLE 8.5 A summary of the process of translation

? 6   What types of bonds other than peptide bonds are present in protein?

DNA REPLICATION

In 1958 two scientists called Meselson and Stahl carried out a classic experiment to show how DNA replicates.

Below is a summary of the experiment they did. Figure 8.5 shows the results they obtained and the conclusions they drew.

- Bacteria were grown in a medium with nucleotides containing a heavy isotope of nitrogen, ^{15}N.
- Some of the bacteria were broken up and centrifuged. The position of the DNA (containing heavy nucleotides) in the sample tube was noted.
- This parental generation of bacteria was then allowed to reproduce in a medium of nucleotides containing a light isotope of nitrogen, ^{14}N.
- Some of this first generation of bacteria were broken up and centrifuged.
- The position of the DNA in the sample tube was found to be different (see Figure 8.5).
- This procedure was repeated to produce a second generation.

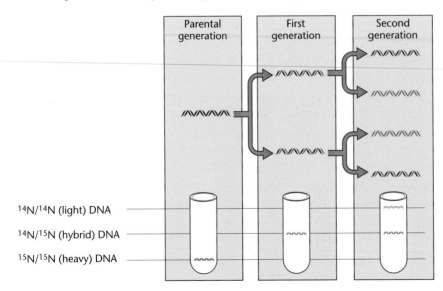

FIGURE 8.5 The results of Meselson and Stahl's experiment

◆ The evidence from this experiment suggests that one of the strands of DNA remains intact, i.e. is conserved, whilst the second strand is made using the available nucleotides.

◆ As only half of the DNA is conserved, this process is known as semi-conservative replication.

---

7 Explain the difference between replication and transcription.

8 What molecule in DNA contains the element nitrogen?

9 Look at Figure 8.5. Draw what you would expect to see if a third generation of bacteria, grown on a medium containing nucleotides made of a light isotope of nitrogen, were broken up and centrifuged.

EXAMINER'S TIP

◆ It is important to know that when $^{14}N/^{14}N$ DNA is centrifuged, the molecule will end up higher in the test tube than the heavier forms.

◆ If a further generation is produced, the lighter band will get thicker but no new bands will appear.

◆ Other experiments may be described. Read the information given in a question carefully and use your knowledge to explain what has happened.

How does DNA affect the phenotype?

You inherit a unique set of DNA molecules from your parents, which give you unique characteristics. Follow the steps below to see how this happens.

◆ A sequence of bases in DNA enables a cell to make a protein.

◆ Many proteins are enzymes.

◆ Enzymes control chemical reactions in a cell.

◆ The sum of the chemical reactions is known as cellular metabolism.

◆ The effect of an enzyme's activity often results in the formation of a product.

◆ This product can often be observed or detected.

◆ This detectable feature is called a phenotype e.g. eye colour, height.

◆ Therefore, each phenotype is linked back to the DNA you inherit.

WORKED EXAM QUESTION

1 a) **Table 1** shows the percentage of different bases in DNA from different organisms.

Source of DNA	Adenine %	Guanine %	Thymine %	Cytosine %
Human	30	20	30	20
Rat	28	22	28	22
Yeast	31	19	31	19
Turtle	28	22	28	22
E. coli	24			
Salmon	29	21	29	21
Sea urchin	33	17	33	17

(i)   What information about the ratios of the different bases in DNA can you work out from the table? *(2 marks)*

The percentage of adenine is the same as the percentage of thymine, 30 to 30 and 28 to 28 so a ratio of 1 to 1.

> There are two marks for this question and the stem clearly states 'ratios' of the 'different bases'. The pattern the candidate has described is correct but there is also a similar relationship for guanine and cytosine which has not been mentioned.

(ii) Give the results that you would expect for DNA from the *E. coli* bacterium. Explain how you arrived at your answer. *(3 marks)*

Guanine **26**   Thymine **24**   Cytosine **26**

Explanation

The percentage of thymine is the same as the percentage of adenine which accounts for 48 percent of the bases. As the ratio of the other bases is also 1 to 1, of the 52 percent left each must be 26 percent.

> A really logical answer, which shows that the candidate did understand that there are two pairs of complementary bases and that as the values given in the table are percentages, the total must come to 100.

(iii) Turtles have the same percentages of the four different bases as rats. Explain why they can still be very different animals. *(1 mark)*

The DNA may contain the same ratio of bases but can code for very different proteins. The turtle DNA codes for turtle protein and the rat DNA codes for rat protein.

> The candidate has missed the point, and although this seems a plausible answer it fails to emphasise that the sequence of the bases/nucleotides determines the proteins synthesised, not the amounts of the bases present.

b) **Table 2** shows the percentage of different bases in the DNA from a virus.

Adenine %	Guanine %	Thymine %	Cytosine %
25	24	33	18

(i) Describe how the ratios of the different bases in this virus differ from those in **Table 1**. *(1 mark)*

There does not seem to be any link between the bases.

The answer is not clear. What does the candidate mean by 'link'? As this is DNA, experience suggests that bases are paired; if that were the case, the percentages of adenine and thymine or cytosine and guanine would be the same. That is the point that should have been emphasised.

(ii) The structure of the DNA in this virus is not the same as DNA in other organisms. Suggest what this difference in DNA structure might be. *(1 mark)*

DNA is not found in chromosomes in a virus.

This candidate has not used the information given in Table 2, which suggests that as the bases do not appear in equal percentages, they are not paired. So perhaps the DNA is not made up of two strands of nucleotides, perhaps it is a single strand. Do not confuse DNA and chromosomes, they are not the same.

c) Describe how proteins are synthesised using the DNA code. *(7 marks)*

DNA is found in the nucleus. It is a very large molecule and is too big to leave. Proteins are made in the cytoplasm in organelles called ribosomes which are associated with the endoplasmic reticulum to form rough endoplasmic reticulum. Messenger RNA is made in the nucleus from DNA and travels through pores in the nuclear membrane to get to the cytoplasm and the ribosomes. Messenger RNA or mRNA as it is known enters the ribosome and a small section or codon is used. This consists of three nucleotides and is the code for one amino acid. Another type of RNA called transfer RNA or tRNA combines with an amino acid and brings it to the ribosome. The tRNA has a complementary set of nucleotides called the anticodon which matches with the codon of the mRNA. The mRNA is made up of a line of codons and this determines the order in which amino acids are joined to make a polypeptide. The amino acids are joined by peptide bonds.

There is a lot of material here that is padding and will not gain credit. The candidate has not concentrated on the question, which expects detail of both transcription and translation. Transcription is incorrectly described as making mRNA from DNA with no further information in terms of the separation of the double helix, the importance of the sense strand, complementary pairing of the RNA and DNA nucleotides or the joining of those nucleotides. It is also much too vague with only a few statements at AS level.

(AQA 2001)

EXAMINATION QUESTION

1   DNA is made up of two polynucleotide strands, the sense strand and the antisense strand. Messenger RNA is transcribed from the DNA sense strand, which contains the genetic code.

   a) The graph shows the number of bases found in the sense strand and the antisense strand of a short piece of DNA, and the mRNA transcribed from it.

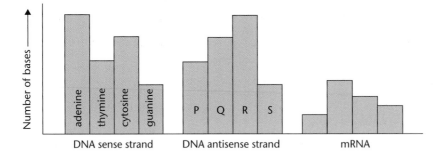

   (i)   Identify the base represented by each of the following letters: **P, Q, R, S**.
         (2 marks)

   (ii) Explain why the total number of bases in the DNA sense strand and the total number of bases in the DNA antisense strand are the same.
         (1 mark)

   (iii) Explain why the total number of bases in the DNA sense strand and the total number of bases in the mRNA are different.   *(1 mark)*

   b) The mRNA has a sequence of 1824 bases. How many amino acids will join to form the polypeptide chain?   *(1 mark)*

   c) Although DNA is double-stranded, only the sense strand determines the specific amino acid sequence of a polypeptide. Suggest a role of the antisense strand.   *(1 mark)*

(AQA 2001)

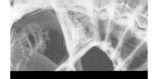

9

The cell cycle

After revising this topic, you should be able to:

▶ interpret diagrams and photographs showing the behaviour of chromosomes during mitosis

▶ show that you have performed simple staining techniques to investigate mitosis in plant material

▶ relate DNA replication and mitosis to the events in the cell cycle

▶ show an understanding of the importance of meiosis in maintaining the chromosome number during sexual reproduction

▶ compare the genetic identity of cells produced by mitosis with those produced by meiosis.

Chromosome copying

Each chromosome has a double-stranded DNA helix surrounded by proteins (called histones). Chapter 8 summarises how DNA is copied by semi-conservative replication. Figure 9.1 relates this to the appearance of a chromosome during cell division. Notice that:

◆ each chromosome has a 'waist', called the centromere, somewhere along its length
◆ after replication, the two copies of DNA (surrounded by histones) are held together at their joint centromere. In this state, each copy of the original chromosome is referred to as a chromatid. They become chromosomes again when they are separated.

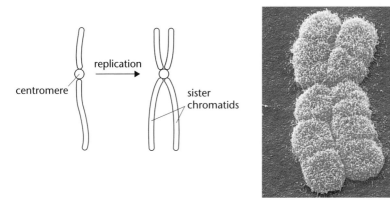

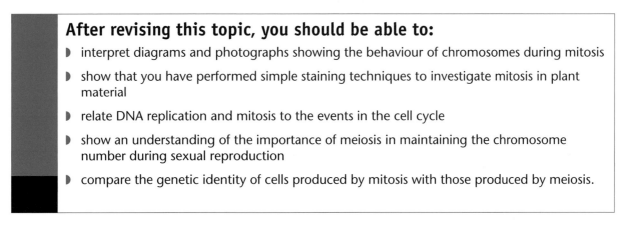

FIGURE 9.1 The appearance of a chromosome during cell division

Cell division

The two types of cell division that occur in eukaryotic cells are:

1 mitosis – which occurs during growth and the replacement of cells
2 meiosis – which occurs at one stage in a life cycle that involves sexual reproduction.

You need to be familiar with the processes of only one of these types of cell division – mitosis. In preparing for your AS examination, do not waste time by revising the processes involved in meiosis. Before revising the processes in mitosis, it is a good idea to ensure that you understand the overview of mitosis in comparison with meiosis shown in Table 9.1.

Feature	Mitosis	Meiosis
Biological processes in which cell division is involved	Production of new cells from old, e.g. during growth, repair of damaged cells or replacement of lost cells.	Involved at only one stage of a life cycle that includes sexual reproduction.
Number of chromosomes in daughter cells compared with parent cell	Daughter cells have exactly the same number of chromosomes as the dividing parent cell – they have a copy of every chromosome that was in the parent nucleus.	Daughter cells have half the number of chromosomes of the dividing parent cell – they have a copy of only one chromosome from each homologous pair. They are referred to as haploid.
Genetic identity of daughter cells compared with parent cell	All daughter cells are genetically identical to the dividing parent cell and to each other. If mitosis has produced large numbers of daughter cells, they form a clone.	Daughter cells are genetically different from the parent cell and from each other. (This idea is developed in your A2 course – do not worry about it here.)

TABLE 9.1 This comparison of mitosis and meiosis is sufficient for your needs at AS level.

1 Give the meaning of the following terms which are emboldened in Table 9.1:
a) clone, b) homologous pair (of chromosomes), c) haploid.

2 Human cells have 46 chromosomes. How many chromosomes will there be in:
a) a lymphocyte cell produced in response to exposure to an antigen,
b) a sperm cell?

EXAMINER'S TIP

Meiosis does not always occur during the formation of sex cells (gametes), as it does in humans. Be prepared for examiners to use unfamiliar life cycles to test your understanding of cell division. Remember to look for where the number of chromosomes has halved: that is where meiosis has occurred. Otherwise, cell division is by mitosis, wherever it occurs in the life cycle.

MITOSIS

You need to be familiar with the processes involved in this type of cell division. This means that you should be able to:
- describe, in continuous prose, the behaviour of the chromosomes and the role of the spindle
- identify drawings or photographs of mitosis and relate their appearance to the behaviour of the chromosomes.

Although mitosis is a continuous process, it is described as a series of stages, depending on the appearance of the chromosomes. These stages are prophase, metaphase, anaphase, telophase and interphase. Table 9.2 summarises in words and in pictures the events that occur during each stage.

Stage	Appearance	Events in cell
Prophase		◆ Chromosomes coil, becoming shorter and fatter (as a result, we can now see them using an optical microscope). ◆ Nuclear envelope disappears. ◆ Network of protein fibres forms a spindle in the centre of the cell.
Metaphase		◆ Chromosomes line up independently on equator of spindle. ◆ Protein fibres of spindle attach to centromere holding each pair of sister chromatids together.
Anaphase		◆ Centromeres divide. ◆ Spindle fibres contract and pull sister chromatids to opposite ends of spindle. As soon as they separate, the sister chromatids are called chromosomes again.
Telophase		◆ Two groups of chromosomes form at opposite ends of spindle – each contains one member of each pair of sister chromatids. ◆ Nuclear envelopes form around each group of chromosomes to form new nuclei. ◆ Chromosomes uncoil to become long and thin. ◆ In most cases, cytoplasm divides to form two new cells.
Interphase		◆ The new cells produce more cytoplasm and carry out their normal functions. ◆ At some stage, cells replicate their DNA.

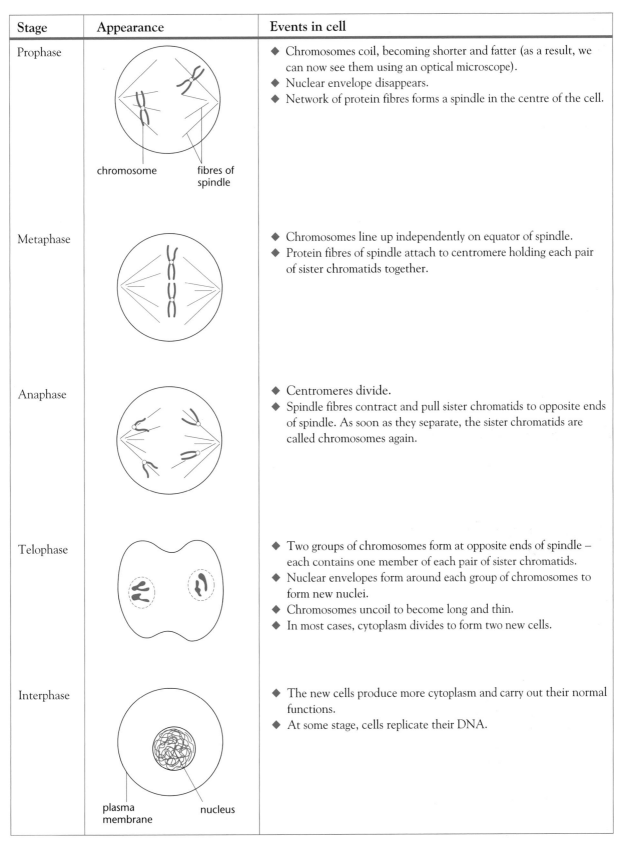

TABLE 9.2 A summary of mitosis

EXAMINER'S TIP

If you are given artwork showing the stages of mitosis in an examination, identify individual stages of mitosis in the following way:

1 Look for the stage at which two bundles of chromosomes lie on each side of the equator of the spindle.
   – If you can see individual chromosomes, this is anaphase.
   – If you cannot see individual chromosomes clearly and the chromosomes are at opposite ends of the cell, this is telophase.
2 Then look for the stage at which all the chromosomes appear in the middle of the cell – this is metaphase.
3 Now look for a stage in which parts of the chromosomes appear visible but they are scattered about the cell – this is prophase.

? 3 How would you recognise a cell that was in the interphase stage of mitosis?

Studying mitosis in plant tissue

You will probably use the growing tips of a plant root to investigate mitosis. Table 9.3 summarises and explains the steps to follow in using root tips from onion bulbs.

Step in procedure	Explanation
1 Cut about 3 mm from the tip of a growing root and put it in a watch glass.	Mitosis occurs only at the growing tip. (Behind this, cells are elongating and differentiating.)
2 Add 1 mol dm^{-3} acetic acid to the watch glass. Warm the acid but do not let it boil.	This softens the plant tissue so that cells separate.
3 Put the root tip on to a glass slide using forceps.	We always mount tissue on a glass slide before examining it under a microscope.
4 Add a few drops of aceto-orcein stain to the tissues on the slide.	Aceto-orcein stains chromosomes, making them easier to see.
5 Gently break up the root tip using a mounted needle. Try to separate the cells rather than stir them up into a ball.	This separates cells so that they are not lying on top of each other when you view them.
6 Lower a coverslip over the stained tissue. Put a piece of filter paper over the coverslip and *gently* press down.	This produces a single layer of cells, making them easier to see.
7 Examine the tissue using an optical microscope.	You should now see individual cells with their chromosomes stained a reddish-purple colour. If you have done this well, the appearance of your chromosome squash should resemble Figure 9.2.

TABLE 9.3 The stages in producing a root tip squash. This process enables you to see cells that were dividing by mitosis.

FIGURE 9.2 A root tip squash

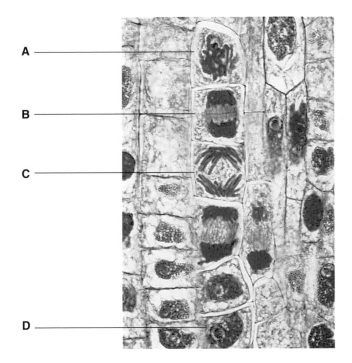

A

B

C

D

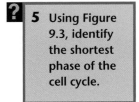

4 Figure 9.2 shows a successful root tip squash. Place the labelled cells in the correct order of the stages of mitosis, starting with the earliest stage.

THE CELL CYCLE

Cells spend most of their time carrying out their normal functions. In those that can divide, only a small part of their time is spent dividing. Figure 9.3 shows mitosis in the context of a cell cycle. The activities of the cell in each phase of this cycle are:

◆ G1 – cell growth involves production of new cell organelles, formation of tRNA and mRNA and protein production

◆ S – semi-conservative replication of DNA occurs

◆ G2 – spindle protein is formed, mitochondria and chloroplasts (if present) divide

◆ M – mitosis occurs (see Table 9.2).

5 Using Figure 9.3, identify the shortest phase of the cell cycle.

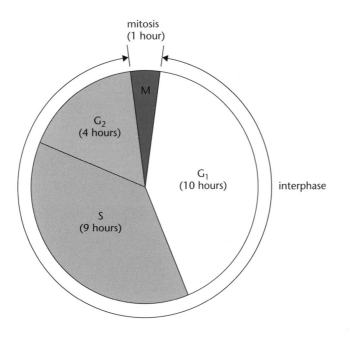

FIGURE 9.3 The cell cycle

MEIOSIS

The cells throughout your body are diploid. This means that they contain two copies of every chromosome.

◆ These two copies form an homologous pair of chromosomes – one of the pair originally came from your mother's egg cell and the other originally came from your father's sperm cell.

◆ The members of each pair carry genes controlling the same characteristics in the same order along the chromosome.

◆ Unlike mitosis, meiosis produces new cells with only one chromosome from each homologous pair.

You do not need to recall the stages of meiosis in your AS course – this comes later in the A2 course. For now, all you need to remember is the following:

◆ Meiosis produces cells with half the number of chromosomes of the parent cell (normally, haploid cells from diploid cells). This is not a random halving of chromosome number. Each daughter cell contains one of the chromosomes from each homologous pair.

◆ Meiosis precedes the fertilisation of gametes, so that fertilisation produces a diploid cell.

Figure 9.4 shows two different life cycles. You will be familiar with one – it is your own. You will be less familiar with the second and do not need to learn it.

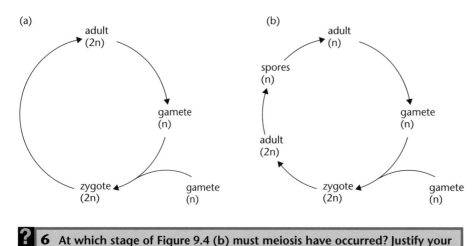

FIGURE 9.4 (a) The human life cycle. (b) A plant life cycle.

? **6** At which stage of Figure 9.4 (b) must meiosis have occurred? Justify your answer.

WORKED EXAM QUESTION

1 a) The diagram shows four stages of mitosis in an animal cell. The stages are not in the correct sequence.

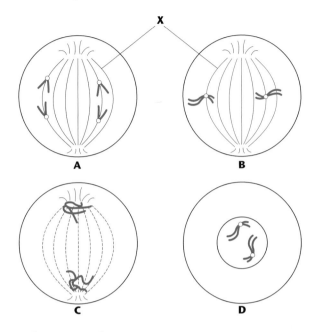

(i) List the stages in the correct sequence. *(1 mark)*

D, B, A, C

The candidate has the correct sequence and gains the mark. Any error in the sequence would result in no mark being awarded.

(ii) What is the function of structure **X**? *(1 mark)*

Structure X is a fibre of the spindle. It attaches to the centromere and pulls apart the sister chromatids during anaphase.

This answer contains more than is needed to gain one mark. The candidate was not asked to name structure X. Information about attaching to the spindle would be fine for a second mark in a two-mark question, but all that was rewarded was information about pulling chromosomes/chromatids apart. Repeated through an examination, this candidate would waste time giving answers for which no further reward can be given.

b) The graph shows how the amount of DNA in a cell varies during the cell cycle.

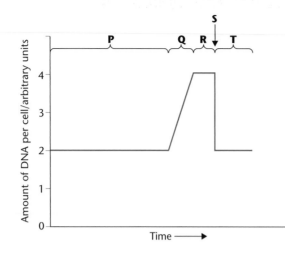

Radioactive thymine was supplied to the cells of some growing tissue. The radioactivity of the nuclei increased during period **Q**.

(i) Explain why the radioactivity of the nuclei increased during period **Q**.

(*3 marks*)

The nucleus contains the chromosomes, which are made of DNA. During period Q the chromosomes replicated and so the amount of radioactivity increased.

The candidate is thinking along the right lines and gains two marks. To gain the third mark, he would have to make the link between thymine and DNA, i.e. stating that thymine is a component of DNA.

(ii) Explain why an increase in the amount of DNA is important in the cell cycle.

(*1 mark*)

The parent cell makes daughter cells that are genetically identical to itself. These cells form a clone. If it did not make an exact copy of every chromosome, it could not do this and the daughter cells would not contain the correct number of chromosomes and so would be a mutant.

In this answer, the candidate has given a lot of information, presumably hoping that something will gain the one mark. In this case it does, but there is always the risk of giving conflicting information where an incorrect statement cancels out a correct statement. The mark was awarded for identifying that the parent cell produces identical daughter cells.

(*AQA 2003*)

EXAMINATION QUESTION

1   The diagram shows some of the different stages in the cell cycle.

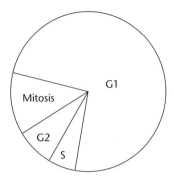

a) There are 20 units of DNA in a cell during stage G2. Give the number of units of DNA you would expect to find in this cell

(i)   at prophase of mitosis;

(ii) in one of the daughter cells produced at the end of mitosis;

(iii) during stage G1.   (*3 marks*)

b) Vincristine is a drug used in the treatment of cancer. It prevents spindle formation during mitosis.

(i)   Explain how treatment with vincristine will affect the behaviour of chromosomes during mitosis.   (*2 marks*)

(ii) People who are given vincristine to treat cancer have a reduced number of red blood cells. Suggest a reason for this.   (*1 mark*)

(AQA 2002)

10 Gene technology

After revising this topic, you should be able to:

▶ explain the term recombinant DNA technology

▶ show an understanding of how a desired gene can be isolated, inserted into a vector and transferred to a population of fast-growing cells

▶ explain the use of reverse transcriptase, restriction endonuclease and ligase enzymes in the above processes

▶ show awareness of moral and ethical issues relating to the use of recombinant DNA technology.

Recombinant DNA technology

Gene technology is often called recombinant DNA technology because it involves putting a gene from one organism into the DNA of another organism, the host.

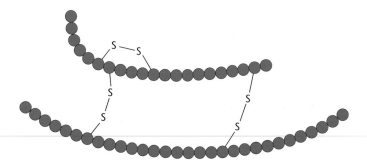

FIGURE 10.1 Human insulin is a protein made of 51 amino acids.

Figure 10.1 shows that insulin is a small protein. Like all proteins, insulin is made of amino acids whose order is determined by the sequence of bases in a gene. Figure 10.2 shows how the gene for human insulin can be inserted into host bacterial cells and used to manufacture human insulin.

ISOLATING THE GENE CODING FOR THE REQUIRED PROTEIN

Since we know the amino acid sequence of insulin, we could make the specific DNA–nucleotide sequence of its two polypeptides.

Table 10.1 shows this and other methods that can be used to isolate the gene for human insulin as well as those controlling the production of more complex proteins.

1 Which of the following aspects of protein structure is shown in Figure 10.1: primary structure, secondary structure, tertiary structure or quaternary structure?

Method of isolating gene	Description
Work backwards from protein	Make DNA sequence from knowledge of amino acid sequence of protein. Mix appropriate DNA nucleotides together with DNA polymerase which controls the joining of the nucleotides.
Use mRNA that codes for the protein	By mixing mRNA for our protein with DNA nucleotides and the enzyme reverse transcriptase, we can make DNA with the desired base sequence. mRNA + DNA nucleotides $\xrightarrow{\text{reverse transcriptase}}$ DNA In the case of insulin, we would expect large amounts of mRNA for insulin to be present in cells in the islets of Langerhans in the pancreas. It might be easier to collect these than to find the insulin gene among the 46 human chromosomes.
Using DNA probes	A DNA probe is a short, single strand of DNA that carries part of the base sequence of the gene we are looking for. In suitable conditions, a DNA probe will attach to the complementary base sequence in the gene we wish to isolate. Because it has a radioactive or fluorescent marker attached to it, we can find the probe and, consequently, the target gene.
Cutting the gene out of its DNA chain	A restriction endonuclease is an enzyme that cuts DNA strands apart at a specific nucleotide sequence – called the recognition sequence of the enzyme. Figure 10.3 shows the action of the restriction enzyme used to cut the insulin gene from a human chromosome.

TABLE 10.1 Methods for isolating genes

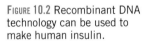

FIGURE 10.2 Recombinant DNA technology can be used to make human insulin.

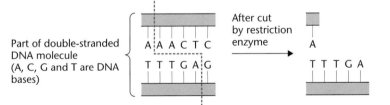

FIGURE 10.3 A restriction enzyme cuts the human insulin gene at a specific recognition sequence. The cut fragments have ends with unpaired bases. Because the unpaired bases will readily bond to other DNA strands with the complementary base sequence, they are known as 'sticky ends'.

2 Suggest why it might be easier to use mRNA that codes for insulin than to cut the insulin gene from its DNA.

3 Name the role of the following enzymes: a) reverse transcriptase, b) DNA polymerase, c) restriction enzyme.

4 Give the recognition sequence of the restriction enzyme that cuts the gene for insulin in Figure 10.3.

PUTTING THE GENE INTO A VECTOR

In the case of the gene for human insulin, we use:

◆ the bacterium *Escherichia coli* as the fast-growing host cells that will make the insulin
◆ a plasmid as the vector for getting the human insulin gene into cells of *E. coli*.

Table 10.2 summarises the stages by which an isolated gene for human insulin is spliced into a plasmid.

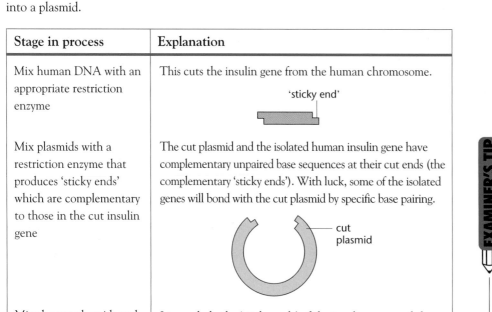

Stage in process	Explanation
Mix human DNA with an appropriate restriction enzyme	This cuts the insulin gene from the human chromosome.
Mix plasmids with a restriction enzyme that produces 'sticky ends' which are complementary to those in the cut insulin gene	The cut plasmid and the isolated human insulin gene have complementary unpaired base sequences at their cut ends (the complementary 'sticky ends'). With luck, some of the isolated genes will bond with the cut plasmid by specific base pairing.
Mix the cut plasmids and the isolated human insulin genes with a ligase enzyme	Ligases help the 'sticky ends' of the insulin gene and the 'sticky ends' of the plasmids to join together. Where this occurs successfully, we have produced recombinant DNA. 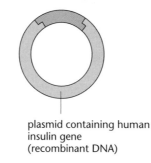

TABLE 10.2 Stages in splicing a human gene into a vector, such as a plasmid

INTRODUCTION OF RECOMBINANT DNA INTO A HOST CELL

Mixing the treated plasmids and the host bacterial cells together in the right conditions produces some bacterial cells that contain plasmids with recombinant DNA.

◆ This process is very hit and miss. Only about 0.0025% of the treated bacterial cells take up the modified plasmid.
◆ Cells with DNA from another organism (humans in this case) are called transgenic organisms.
◆ Once we have identified these transgenic bacteria, we can culture them and they will produce the desired product (in this case, human insulin).

> **? 5** What is a plasmid?

> **EXAMINER'S TIP**
> The staggered cuts of the isolated gene and of the plasmid must have complementary base sequences exposed. You must make this clear in an examination. You can describe this in one of two ways: the gene and the plasmid are cut with the *same* restriction enzyme, or they are cut with restriction enzymes that produce *complementary* 'sticky ends'.

> **? 6** What is a transgenic organism?

FINDING THE GENETICALLY MODIFIED BACTERIA

We do this by using other genes whose effects we can see. Table 10.3 shows a commonly used method, involving genes for resistance to antibiotics.

Stage in process	Explanation
A plasmid is chosen that contains two genes, each giving rise to resistance to a different antibiotic. The human gene is inserted into the middle of one of these genes for antibiotic resistance.	The undisturbed gene will confer resistance to antibiotic A in any bacterium that has taken up the modified plasmid. The gene into which the human gene has been spliced will no longer function, so bacteria will be susceptible to antibiotic B.
The bacteria are grown on a solid medium containing antibiotic A.	Any bacterial colony that grows on the medium must have taken up a plasmid with the gene for resistance to antibiotic A. These are the ones we are interested in.
A velvet fabric is used to take an imprint of the colonies of bacteria and to transfer them to a new solid medium containing antibiotic B.	The position of each bacterial colony is known from its position on the first solid medium.
After incubation, the position of any new colonies is compared with those on the first solid medium.	Any colony growing on the second medium is resistant to antibiotic B, so its gene for resistance to antibiotic B must be intact – it does not have the human insulin gene. We reject these. Any colony from the first medium that is absent from the second medium does not have a functional gene for resistance to antibiotic B. This must have taken up the human insulin gene. We remove this colony from its known position on the first medium and culture it. This can be used to give us a pure culture of cells that produce human insulin.

TABLE 10.3 Stages in the identification of bacterial cells that have taken up the desired gene

> **7** Explain why the colony of bacteria labelled X in Table 10.3:
> a) shows that these bacteria must have taken up a plasmid,
> b) shows that these bacteria must contain the gene for human insulin.

CULTURING THE HOST CELLS

A pure culture of cells that contain the desired gene is grown in large tanks called fermenters. As the modified bacteria grow in culture, they produce the target human protein, e.g. insulin. At suitable intervals, the culture solution is drawn off and the human protein is extracted from it by downstream processing. There is more about fermenters and downstream processing in Chapter 7.

EXAMINER'S TIP

Examination questions often ask for a brief account of genetic engineering. The mark scheme is fairly constant from year to year and contains the following marking points:

◆ Locate desired gene/use of DNA probe.
◆ Cut gene from DNA using a restriction enzyme.
◆ Cut plasmid using same restriction enzyme/restriction enzyme that produces complementary 'sticky ends'.
◆ Join isolated gene and plasmid using ligase enzyme.
◆ Mix plasmids with bacteria under appropriate conditions.
◆ Determine which bacteria have taken up plasmid with desired gene.
◆ Description of how these bacteria can be identified.
◆ Culture bacteria and harvest product formed.

MORAL AND ETHICAL ISSUES ASSOCIATED WITH RECOMBINANT DNA TECHNOLOGY

The introduction of human insulin genes into bacteria is one example of recombinant DNA technology. Other uses include:
◆ inserting human genes into cattle or sheep so that they produce a human protein in their milk
◆ inserting genes for resistance to pests, or to herbicides, into crop plants
◆ inserting growth-controlling genes into farmed animals
◆ inserting genes for tolerance to cold or dry conditions into crop plants.

The use of human insulin produced by recombinant DNA technology has benefited sufferers of insulin-dependent diabetes. These people need to inject themselves with insulin.
◆ Since insulin from cattle or pigs is slightly different from human insulin, lifelong use of these hormones caused complications among insulin users, including blindness. Human insulin does not cause these complications.
◆ Initially, users of human insulin found that they had no warning of a fall in their blood glucose concentration, which they had experienced using insulin from cattle or pigs. This was an initial problem of using human insulin.

However, there are other issues involved in the use of recombinant DNA technology which cause concern. Some of these are summarised in Table 10.4.

Use of DNA technology	Arguments for use	Arguments against use
A new gene is inserted into the transgenic organism.	◆ Yield of organism increases, e.g. by faster growth. ◆ Organism produces desired protein product.	New gene might behave differently in the genotype of the transgenic organism from the way it did in the source organism. We cannot predict what this new effect might be.
Bacteria are produced that contain genes for resistance to antibiotics.	◆ The resistance to antibiotic can be used to detect cells that have taken up the desired gene (see Table 10.3). ◆ Cells that have taken up the desired gene can be used to make a desired product, such as human insulin.	Mutation, or 'escape', of these microorganisms might result in antibiotic-resistant pathogens, which could cause epidemics.
Useful transgenic animals are produced.	◆ The growth rate, and hence productivity, of transgenic animals can be increased. ◆ The milk of transgenic animals could be a source of a desirable protein in a form that is easy to harvest.	Genes often have more than one effect. The full effect of the new gene in the transgenic organism is not known and it could have a harmful effect on the host animal.
Transgenic crop plants are produced.	The yield of the crop increases, e.g. through resistance to herbicides or pesticides.	◆ See row above. ◆ Gene might spread to other species of plants, e.g. weeds, and give them resistance to herbicides or pesticides.
Transgenic organisms are kept in the 'field'.	Large numbers of the organism can be grown, e.g. transgenic crop plants or populations of transgenic fish.	◆ Transgenic populations might compete successfully with natural populations and replace them through natural selection. ◆ New populations from above might change food chains, causing a shift in the stability of biological communities.
Transgenic food organisms might be used to control the population of mammal populations that are pests.	Populations might be controlled more cheaply, more humanely or more effectively than the use of poisoning, hunting or shooting.	◆ Other species of animal might eat the transgenic food organism and be killed. ◆ The effect on non-target mammalian populations might disrupt food chains.
Human genes might be transferred to animals such as cattle or pigs.	Desirable human proteins can be harvested and used to help humans suffering from illnesses, such as insulin-dependent diabetes.	Many religious communities hold strong views about certain animals. Cows are sacred to Hindus and pigs are unacceptable to Jews and Muslims.
Genes can be inserted into humans to counteract the effect of a genetic disorder.	Sufferers of genetic disorders, such as cystic fibrosis, can be 'cured'.	◆ Money spent on this technique is diverted from other medical uses which might benefit more people. ◆ Many people are concerned that the genetic modification of humans is similar to eugenics programmes which have been used throughout history to eradicate less powerful ethnic groups.

TABLE 10.4 There are many issues raised by the use of DNA recombinant technology – these are some of them.

109

Candidates often make emotional statements about moral or ethical considerations of recombinant DNA technology. However strongly you hold views about the potential uses of this technique, you should attempt to maintain a balanced view for the purposes of an examination. Often questions will ask for arguments for and against, so you must be able to give both in order to gain full marks.

The following commonly seen responses **do not** gain marks.

◆ 'Playing at God'.

This expression has no scientific meaning. The existence, or activities, of God are matters of faith and cannot be investigated scientifically.

◆ 'Humans have no right to interfere with organisms' or 'Animals have rights too'.

What we have the right to do and whether animals have rights are moral issues. These rights are decided by each society and might be different in different societies or over time in any one society.

Bear in mind that humans have 'interfered' with organisms, using artificial selection, since the cultivation of crop plants and domestication of animals began several thousand years ago.

WORKED EXAM QUESTION

1   Read the following passage.

Large numbers of possums in New Zealand are eating crops and spreading disease between cattle. The use of shotguns and poisons by farmers has not greatly reduced possum numbers. A better solution to the possum problem may have to be found. A crop of carrots has been genetically modified to produce a 'sterility protein'. This sterility protein prevents possums producing offspring.   5

First, scientists identified the gene that codes for this sterility protein. Several copies of the sterility gene were cut out from long sections of DNA using a special enzyme. The same enzyme was also used to cut open plasmids which had been removed from bacterial cells. A different enzyme joined together the 'sticky ends' of a plasmid and of a sterility gene to produce a recombinant plasmid.   10

The scientists then tried to put these plasmids back into bacteria. Each plasmid also contained a gene, giving resistance to an antibiotic which normally kills bacteria. Because of this resistance gene, the scientists could identify bacteria containing the sterility gene and isolate them from bacteria which had not taken up this gene. Finally, carrot seedlings were sprayed with bacterial cells. The plasmids entered the carrot seedlings and carried   15 copies of the sterility gene into the DNA of carrot cells.

The genetically modified crop will be harvested and the carrots scattered across land populated by possums.

Use information from the passage and your own knowledge to answer the following questions.

a)   Name the type of enzyme used to:

  (i) cut out the sterility gene (line 6);   *(1 mark)*

     Restriction enzyme.

> The candidate has correctly recalled this term.

  (ii) join together the plasmid and the sterility gene (line 9).   *(1 mark)*

     Ligase.

> Again, a correct term has been recalled.

b)   Explain the meaning of the term 'sticky ends' (line 9).   *(2 marks)*

   Sticky ends are when DNA has been cut to leave unpaired bases at its ends. They are sticky because they will join with strands that have the same base sequence.

> The first mark has been gained for describing unpaired bases. The candidate has made an incorrect statement in the second sentence: the DNA pairs with bases with a complementary sequence, not the same sequence.

c) In this procedure, the bacterial plasmids acted as vectors. Explain the function of a vector. *(2 marks)*

It carries the gene into the bacterial cell.

Both marks gained here: carries gene/DNA; into target cell.

d) Explain how the presence of the antibiotic resistance gene allowed scientists to identify and isolate the bacteria which contain the sterility gene (lines 11–12). *(3 marks)*

A bacterial cell grows on a medium to produce a colony which we can then see. If a bacterium did not take up a plasmid, it will not have the gene for antibiotic resistance and so will not grow on a medium containing antibiotic. This means that any bacteria that produce a colony must have the plasmid and so they must have the sterility protein gene as well.

This is a good answer and gains three marks.

e) Explain the arguments for and against using genetically modified carrots to reduce the population of possums. *(6 marks)*

This is interfering with the environment which is finely balanced and interfering in this way could upset the balance and have a drastic effect on the environment. The scientists do not know what effect the sterility protein might have and so they should not put it into the environment if they do not know what it will do. Also the carrots might be eaten by farm animals and this might sterilise them. People are likely to destroy the crop if they know that genetically modified crops are being grown there.

The candidate seems to have no grasp of this question and has produced a vague, repetitive answer. In the last sentence, the candidate seems to have suddenly realised the term 'genetically modified' has been used in the question and has wandered off into memories of a tabloid headline. The answer has not given arguments for and against use of the technique, as asked for in the question. Often mark schemes limit the maximum number of marks for each part of the discussion: in this case 3 marks maximum for arguments for, and 3 maximum for arguments against.

Arguments for include: using GM carrots is more effective than shooting/poisoning; using GM carrots is more humane than shooting/poisoning; poisons might kill other useful animals; new method will ensure better yields of crops; new method will prevent the spread of disease from possums to cattle (3 marks maximum).

Arguments against include: plasmid might enter another species; other species of animal might be sterilised; (sterilising other animals) might disrupt food chains; possums might become resistant to the sterility protein; (overuse of antibiotic for this technique) might result in bacteria becoming resistant to this antibiotic (3 marks maximum).

(AQA 2001)

EXAMINATION QUESTION

1   The diagram shows how insulin can be made using genetically modified bacteria.

a) (i)   The human insulin gene is obtained from mRNA, rather than DNA. Suggest why.   *(1 mark)*

   (ii) Name the enzyme used to make a single-stranded DNA copy of the mRNA coding for insulin.   *(1 mark)*

   (iii) The table shows a sequence of bases from the mRNA coding for insulin. Complete the table to show the sequence of bases you would expect in the single-stranded DNA copy.   *(1 mark)*

mRNA base sequence	U	C	A	A	C	C
DNA base sequence						

b)   What is the role of DNA ligase in producing genetically-modified bacteria?   *(1 mark)*

c)   The plasmid contains a marker gene coding for antibiotic resistance. Explain the importance of this marker gene.   *(2 marks)*

(AQA 2001)

11 Immunology and forensic biology

After revising this topic, you should be able to:

▷ define the terms antibody and antigen

▷ understand how B-lymphocytes produce an immune response

▷ explain the ABO blood system

▷ explain the techniques of genetic fingerprinting

▷ explain polymerase chain reaction.

Immunity

Most pathogens do not harm us because they are unable to get through our protective barriers, such as skin. If they do penetrate these barriers and enter the bloodstream, they will encounter our immune response. One defence mechanism that we use is known as the humoral response.

> **? 1** Apart from skin, give two other barriers that will stop the entry of pathogens.

There are a number of specific words that you need to know to understand this topic. These are given in Table 11.1.

Term	Definition
Antigen	A molecule that can trigger an immune response.
Non-self antigen	A molecule found on cells entering your body, e.g. bacteria, that will stimulate an immune response.
Self antigen	A molecule found on the surface of your own cells to which your immune system will not respond. If put into somebody else, it will trigger an immune response.
Antibody	◆ A molecule secreted by B-lymphocytes in response to an antigen. ◆ An antibody is always a protein molecule.

Phagocyte e.g. macrophage	A type of white blood cell that engulfs bacteria.
Lymphocyte	A type of white blood cell that comes in a number of different forms, for example B-lymphocyte and T-lymphocyte.
1 B-lymphocytes (B-cells)	◆ They are highly specific and will only respond to one antigen. ◆ They produce only one type of antibody. ◆ The body must be able to produce many different types of B-lymphocyte to respond to the different types of antigen that enter our body.
2 T-lymphocytes (T-cells)	◆ They are highly specific, each having their own unique receptors. ◆ T-helper lymphocytes are very important as the B-lymphocytes are unable to respond to an antigen without them.

TABLE 11.1 Some terms used in the study of immunology

> **2** All antibodies are proteins. How does this fact help an antibody in its function?

EXAMINER'S TIP

You must learn the definitions from Table 11.1. Examiners will expect you to be able to relate them to unfamiliar situations.

Humoral immunity

This is the body's response to non-self antigens found in the bloodstream. Table 11.2 shows this process step by step.

Stage	Appearance
1 Ingestion The macrophage ingests microorganisms containing a non-self antigen.	

2   Antigen presentation
    The non-self antigens are presented on the
    surface membrane of the macrophage.

3   Clonal selection
    The B-lymphocyte and a T-helper lymphocyte attach
    to the macrophage. The lymphocytes have
    complementary receptors to those of the non-self
    antigen.

4   Clonal expansion
    The T-helper cells stimulate the selected
    B-lymphocyte to divide, producing a clone of
    identical B-lymphocytes.

5   Plasma cells produced
    Some of the cloned B-lymphocytes now differentiate
    into plasma cells. These produce the specific antibody.

6   Memory B-lymphocytes
    Some of the cloned B-lymphocytes remain in the
    lymph node to become memory B-lymphocytes.
    These will be used if the same antigen enters again.

TABLE 11.2 The various steps in the body's response to non-self antigens

Immunological memory

The body meets the same pathogens many times during our lives but due to our ability to 'remember' these contacts, these pathogens do not cause us harm time after time.

Table 11.3 shows the differences between the body's primary response to an antigen and its secondary response.

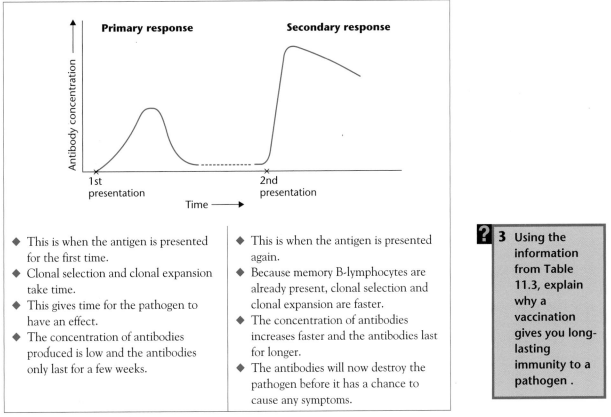

- This is when the antigen is presented for the first time.
- Clonal selection and clonal expansion take time.
- This gives time for the pathogen to have an effect.
- The concentration of antibodies produced is low and the antibodies only last for a few weeks.

- This is when the antigen is presented again.
- Because memory B-lymphocytes are already present, clonal selection and clonal expansion are faster.
- The concentration of antibodies increases faster and the antibodies last for longer.
- The antibodies will now destroy the pathogen before it has a chance to cause any symptoms.

TABLE 11.3 The primary and secondary responses to an antigen

> **3** Using the information from Table 11.3, explain why a vaccination gives you long-lasting immunity to a pathogen .

ABO blood system

Blood groups are determined by the presence of antigens on the membrane of red blood cells. There are two possible antigens that can be present – Antigen A and Antigen B. Combinations of these antigens produce four possible blood groups: **A, B, AB** and **O**. These are shown in Table 11.4.

Blood group	Antigen present on red blood cell
A	Antigen A only
B	Antigen B only
AB	Antigen A and Antigen B
O	Neither Antigen A nor Antigen B

TABLE 11.4 The ABO blood system

In the plasma there are two possible antibodies that can be present:
◆ Antibody a
◆ Antibody b.

It is important to remember that if antigen A meets antibody a, this will cause the red blood cells to agglutinate, or stick together. The complementary antigen and antibody *must* be kept apart. This is also true if antigen B meets antibody b.

Table 11.5 shows the antigens and antibodies found in the blood of particular blood groups.

Blood group	Antigen present	Antibody present in plasma
A	A	Antibody b only
B	B	Antibody a only
AB	A and B	Neither Antibody a nor Antibody b
O	None	Antibody a and Antibody b

TABLE 11.5

WHO CAN GIVE BLOOD TO WHOM?
◆ The person who gives the blood is known as the donor.
◆ The person who receives the blood is known as the recipient.
◆ It is the *antigens of the donor* on the red blood cell that must be taken into account.
◆ It is the *antibodies* that are present in the plasma of the *recipient* which must also be taken into account.

Table 11.6 shows possible acceptable donors and recipients. The letters in bold represent the antigen given by the donor and the letters in *italics* represent the antibody present in the recipient.

		Donor's blood group and antigens present in blood			
		A A	B B	AB A and B	O Neither A nor B
Recipient's blood group and antibodies present in blood	A b	✓	✗	✗	✓
	B a	✗	✓	✗	✓
	AB Neither a nor b	✓	✓	✓	✓
	O a and b	✗	✗	✗	✓

TABLE 11.6 How ABO blood types can be donated. Key: ✓ = Transfusion can take place, no agglutination will occur; ✗ = Transfusion cannot take place, agglutination will occur.

4 Explain why
a) blood group O is described as the universal donor,
b) blood group AB is described as the universal recipient.

EXAMINER'S TIP

Never use the term 'clot' as an alternative to agglutination. Clotting is a totally different biological process and you will not get credit for using the term in this context.

EXAMINER'S TIP

When deciding whether one blood group can be safely given to someone with another blood group, always consider the antigens on the red blood cells of the donor. The antibodies of the donor will be diluted so much that they have no effect.

Genetic fingerprinting

Genetic fingerprinting is a technique often used by forensic scientists to compare the DNA from different sources. It works because DNA from every individual is unique. Table 11.7 describes the steps involved.

Description	Diagram
◆ DNA contains non-coding sequences (introns). ◆ Within these regions, sequences of bases may be repeated. These are called minisatellites. ◆ The number of repeats, and so the size of the minisatellite, varies from one person to another. 1 DNA is isolated from a suitable source such as blood or semen. 2 The DNA is cut into pieces using restriction enzymes. Remember: Some of the pieces will contain the repeated sequences. If there are only a few repeats, the piece of DNA concerned will be small. If there are many repeats, the DNA segment will be large. 3 Electrophoresis is used to separate the pieces of DNA. The pieces of DNA are not visible at this stage. They only become visible when a DNA probe is used (see Stage 4). Remember: The smaller pieces, ones with fewer repeats, will travel further than the longer pieces of DNA, with many repeats.	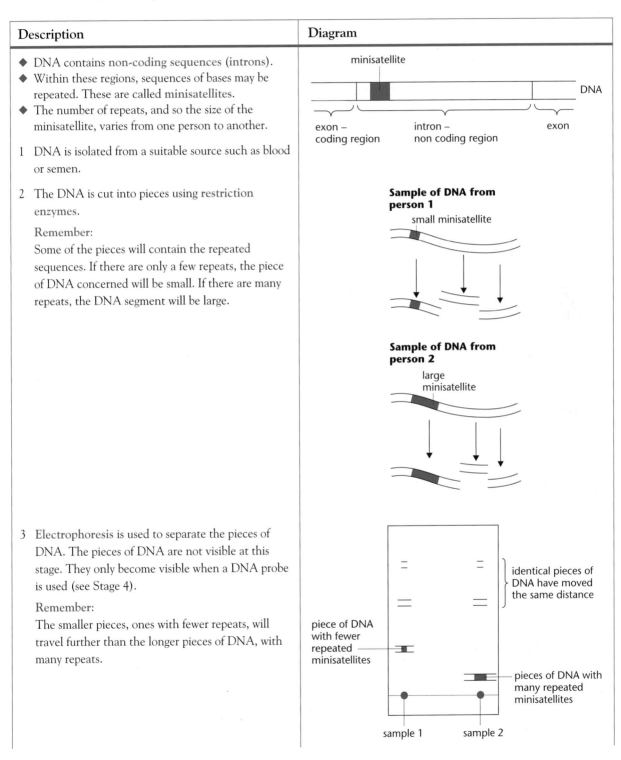

4   The bands of DNA containing the repeated
    sequences are now identified using a DNA probe
    for those sequences.

    Remember:

    A probe is a single strand of DNA nucleotides which
    are complementary to the bases of the minisatellite.
    The probes are radioactive to allow the position of
    the minisatellite to be identified.

5   The position of the bands can be used to compare
    samples from different individuals.

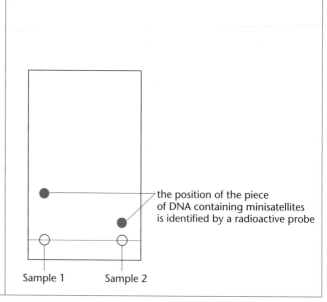

the position of the piece
of DNA containing minisatellites
is identified by a radioactive probe

Sample 1   Sample 2

TABLE 11.7 The steps involved in genetic fingerprinting

> **EXAMINER'S TIP**
>
> Remember that contamination is a problem. Any foreign DNA from sources
> such as bacteria will produce extra bands.

? 5   Why would it be impossible to obtain a sample of DNA from red blood cells?

ANALYSING RESULTS

A woman accused a wealthy footballer of being the father of her child. He said that he
was not. To settle the issue, genetic fingerprinting was carried out on DNA from the
mother, the child and the footballer. The genetic fingerprints from this test are shown in
Figure 11.1.

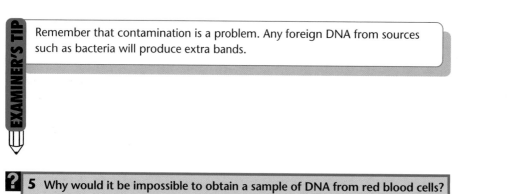

mother

child

footballer

More than one probe
has been used, producing
many bands per sample.

FIGURE 11.1 Using a genetic
fingerprint to determine
paternity

**? 6   Giving reasons
for your
answer, was
the footballer
the father of
the child?**

Polymerase chain reaction

The polymerase chain reaction enables large amounts of DNA to be produced from very small samples.

◆ The original DNA is copied many times.
◆ Large amounts of *identical* DNA are produced for analysis.

Table 11.8 summarises the processes involved.

Method	Diagram
1 DNA is separated into two strands by heating it to a temperature of 95°C.	
2 The sample is cooled and then mixed with DNA nucleotides, DNA polymerase and primers. 3 The primers are single strands of DNA which bind to the ends of the pieces of DNA to be copied.	
4 DNA polymerase now copies each strand of the original DNA by adding nucleotides.	
5 This results in two molecules of DNA identical to the original one. 6 The cycle can be repeated to produce many copies of the original DNA.	

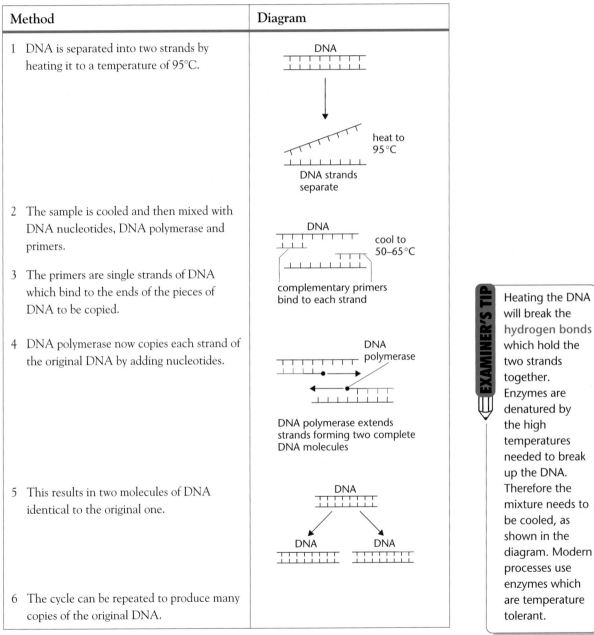

TABLE 11.8 The stages in the polymerase chain reaction

EXAMINER'S TIP

Heating the DNA will break the **hydrogen bonds** which hold the two strands together. Enzymes are denatured by the high temperatures needed to break up the DNA. Therefore the mixture needs to be cooled, as shown in the diagram. Modern processes use enzymes which are temperature tolerant.

WORKED EXAM QUESTION

1   The polymerase chain reaction is a process used to make many copies of a piece of DNA. The process is outlined in the diagram.

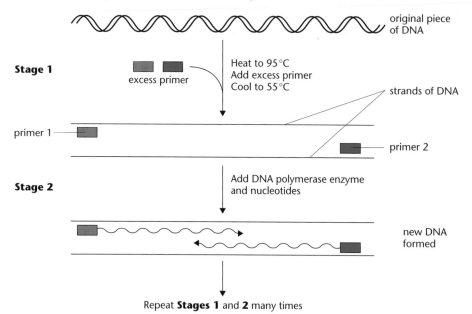

a) Describe and explain what happens to the DNA when it is heated to 95°C in Stage 1.
(2 marks)

The two strands of DNA separate.

> The mark allocation gives a clue as to how much needs to be written. In this case the candidate has given only one point and cannot expect to get both available marks. She has described what has happened but not explained why. Examiners expected the fact that heat breaks the hydrogen bonds holding the two strands together as an explanation.

b) The primer is a short sequence of nucleotides that binds to the DNA.

(i)   Explain why the primer only binds to the DNA at a particular position.
(1 mark)

As the primer is a short sequence of nucleotides it will only join onto a complementary sequence on the DNA.

> The candidate has read the information given in the stem and has used her knowledge about the complementary nature of nucleotides.

(ii) Describe the role of the primer in the polymerase chain reaction. *(1 mark)*

The primer starts the process, without it nothing would happen.

> This answer is too vague. The DNA polymerase enzyme which causes replication will only attach to a double-stranded structure. The primer makes the strand a double structure and therefore gives a starting point for amplification.

(iii) Suggest why two different primers are needed. *(1 mark)*

Because there are two different starting places with different nucleotide sequences.

The sections of DNA are copied in different directions on the two strands. Therefore the complementary primer needs to match a sequence at the beginning of the gene on one strand and a sequence at the end of the gene on the other.

c) It is important that the original DNA sample is not contaminated with any other material, such as bacteria or human skin. Explain why. *(2 marks)*

Cells from bacteria or other sources will contain DNA which will contaminate the pure DNA you want to copy.

Again there is only one point made here. They have identified that there is also DNA in the other cells not that it will also be amplified which is the reason for the serious contamination.

d) The DNA polymerase used comes from a bacterium which lives in hot springs. Explain the advantage of using DNA polymerase from this organism.
(1 mark)

Enzymes are denatured by heat because they are proteins. Using enzymes from these organisms that are used to high temperatures will mean that they are not affected by heating them to 95°C which is the temperature needed in stage 1 to breakdown the original two strands.

This is a very full answer with some very good biology but will not get any more marks. Unfortunately only one mark was allocated to this answer and regardless of the amount written, there are no bonus marks.

(AQA 2001)

EXAMINATION QUESTION

1 **Figure 1** shows the polymerase chain reaction (PCR). This reaction can be used to produce multiple copies of a small amount of DNA, such as might be found in a blood stain at the scene of a crime.

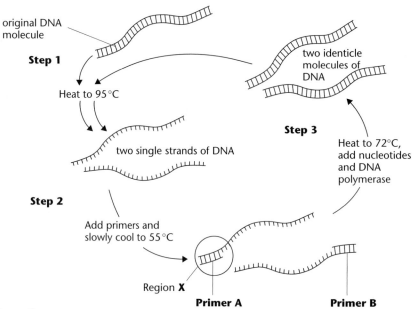

Figure 1

a) What type of bond is broken when the DNA is heated to 95°C in **Step 1**?
(*1 mark*)

b) (i) Why is it necessary to join primer molecules to the single-stranded DNA (**Step 2**) before DNA polymerase is used in **Step 3**? (*1 mark*)

   (ii) **Figure 2** shows region **X** from **Figure 1**. Copy and complete **Figure 2** to show the sequence of bases on **Primer A**. (*1 mark*)

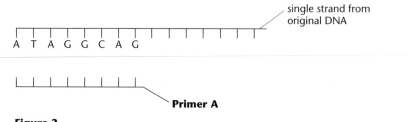

Figure 2

c) Starting with a single molecule of DNA, the polymerase chain reaction was allowed to go through six cycles. How many molecules of DNA would be produced? (*1 mark*)

d) The original DNA must not be contaminated with any other biological material, such as bacteria or human skin cells, when carrying out the polymerase chain reaction. Why is this important? (*2 marks*)

(*AQA 2003*)

12   Crop plants

After revising this topic, you should be able to:

▶ explain how rice, sorghum and maize plants are adapted to the conditions in which they are grown

▶ interpret and explain data showing the effect of light intensity, temperature, carbon dioxide concentration and fertiliser on the productivity of crops

▶ explain the need for the use of fertilisers on crop plants and evaluate environmental issues arising from their use

▶ understand and interpret data concerning the effect of competitors and pests on crop plants

▶ understand the principles of using chemical pesticides, biological agents and integrated systems in controlling pests

▶ evaluate issues involved in different methods of pest control, including environmental issues.

What are crop plants?

A crop plant is one that humans grow for a specific purpose. In cultivating crop plants, humans manipulate the plants themselves and the environments in which they are grown.

Cereals

Cereal crops are grown for their grain (their seed and, sometimes, its surrounding fruit). In Europe, the most commonly eaten cereals are:
- ◆ wheat – used to make bread, pasta and rhoti
- ◆ maize – eaten as corn on the cob and as sweet corn.

These cereals do not grow well throughout the world and other cereals are important crops in different countries. Table 12.1 summarises the adaptations of three crop plants you must know: maize, rice and sorghum. In interpreting this table, it is essential to remember that:
- ◆ plants need water, carbon dioxide (CO_2) and light in order to photosynthesise
- ◆ cells produce ATP by a process called respiration. More ATP is produced if oxygen is present (aerobic respiration) than if oxygen is absent or in limited supply (anaerobic respiration).

Cereal crop	Conditions in which crop plant is adapted to grow	Adaptation(s)
Maize	◆ Hot regions ◆ In these conditions, the pores in the leaves (stomata) through which CO_2 diffuses are often closed. This helps to reduce water loss but leads to a low CO_2 concentration inside the leaf.	◆ Maize plants use a specialised method of photosynthesis (C_4 pathway) which is different from that of temperate plants (C_3 pathway). ◆ The C_4 pathway enables maize to take in CO_2 even at low concentrations, and convert it to a four-carbon compound that can be used as a source of CO_2 by other, photosynthesising cells. ◆ Because the C_4 pathway uses a lot of ATP, it can only occur where light intensities are very high.

Rice	◆ Warm swamps
	◆ These have waterlogged soil with a low oxygen concentration. In these conditions, plant cells respire anaerobically and produce ethanol as a waste product.
		◆ Stems of rice plants have tissues with large air spaces, called aerenchyma (see Figure 12.1). These spaces:
		  – allow oxygen to diffuse easily to cells in the plant's roots, enabling aerobic respiration to continue in them
		  – provide buoyancy, keeping the photosynthesising leaves in the light.
		◆ Ethanol is toxic. However, cells in the roots of rice plants are unusually tolerant of ethanol. This enables anaerobic respiration to continue in cells of rice roots for longer than it would in other plants.

Sorghum	◆ Hot, dry regions ◆ Plants which are adapted to grow in hot, dry conditions are collectively termed xerophytes. Their adaptations increase water uptake and/or reduce water loss.	◆ An extensive root system enables water uptake over an unusually large area of soil. ◆ Very thick, waxy cuticles reduce water loss by evaporation through upper surfaces of leaves (see Figure 12.2). ◆ There are few pores in the leaf (stomata) through which CO_2 diffuses. This reduces water loss through stomata. ◆ The leaf rolls inwards in dry conditions, trapping a layer of moist air. ◆ Adults and embryos are tolerant of high temperatures.

TABLE 12.1 The structural and physiological ways in which maize, rice and sorghum are adapted to different environments

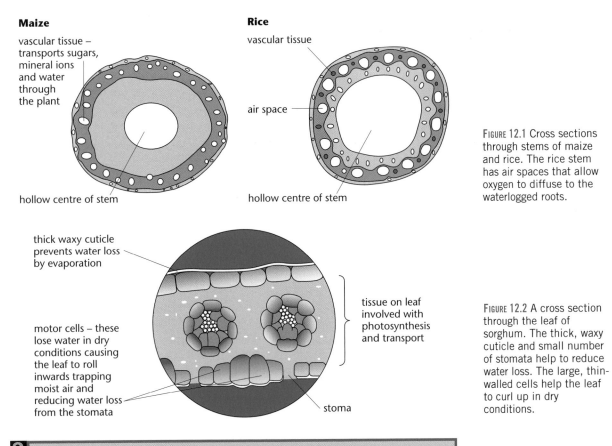

Maize

vascular tissue – transports sugars, mineral ions and water through the plant

hollow centre of stem

Rice

vascular tissue

air space

hollow centre of stem

FIGURE 12.1 Cross sections through stems of maize and rice. The rice stem has air spaces that allow oxygen to diffuse to the waterlogged roots.

thick waxy cuticle prevents water loss by evaporation

motor cells – these lose water in dry conditions causing the leaf to roll inwards trapping moist air and reducing water loss from the stomata

tissue on leaf involved with photosynthesis and transport

stoma

FIGURE 12.2 A cross section through the leaf of sorghum. The thick, waxy cuticle and small number of stomata help to reduce water loss. The large, thin-walled cells help the leaf to curl up in dry conditions.

> **1** Explain why having a small number of sunken stomata helps sorghum to grow in hot, dry conditions.

Controlling the environment in which we cultivate crops

◆ The growth rate of plants is dependent on their rate of photosynthesis.
◆ Humans increase the growth rate of crop plants by controlling the environmental factors needed for photosynthesis to occur.

> **2** What is meant by limiting factor in the context of photosynthesis?

LIMITING FACTORS

The rate of photosynthesis is limited by:

◆ carbon dioxide concentration – carbon dioxide is combined with hydrogen from water to make carbohydrates during photosynthesis
◆ light intensity – light energy is needed to make ATP which is used as an energy source for making carbohydrates from carbon dioxide and water
◆ temperature – the enzymes that speed up the reactions of photosynthesis lose kinetic energy as the temperature falls, and become denatured as the temperature rises.

Table 12.2 summarises the effect of these environmental factors on the rate of photosynthesis.

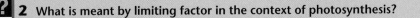

EXAMINER'S TIP

Questions testing your under-standing of limiting factors are common. It is a good idea to make sure you can explain the concept and that you can interpret graphs showing the effect of limiting factors.

Graph	Description	Explanation
(a)	From A to B, an increase in the light intensity results in an increase in the rate of photosynthesis. Between B and C, a further increase in the light intensity does not result in an increase in the rate of photosynthesis.	Light is a limiting factor since the rate of photosynthesis increases when more light is available. Light is no longer a limiting factor. Another environmental factor now limits the rate of photosynthesis.
(b)	For each curve, an increase in the light intensity initially results in an increase in the rate of photosynthesis. Curve D levels out at a higher rate of photosynthesis than curve F. Curve D levels out at a higher rate of photosynthesis than curve E. Curves D, E and F level out at a higher rate of photosynthesis than curve G.	The shape of each curve can be explained as above. The higher rate of photosynthesis in curve D shows that CO_2 concentration was a limiting factor in curve F. Temperature was a limiting factor in curve E. CO_2 concentration and temperature are limiting factors in the level part of curve G.

TABLE 12.2 The effect of environmental factors on the rate of photosynthesis

GLASSHOUSES

Growing a crop in a glasshouse enables growers to control some of the limiting factors and so increase the yield. In a glasshouse:

- artificial lighting can be used to increase light intensity
- paraffin heaters or air pumps can be used to increase carbon dioxide concentration
- paraffin, or electric, heaters can be used to raise the temperature.

Remember that all these methods cost money. It is only worthwhile for a grower to use them if the increased yield results in a greater additional income than the additional expenditure involved. Consequently, a grower will try to achieve an optimum yield rather than a maximum yield.

3 Explain the shape of curve E compared with curve G in part (b) of Table 12.2.

 4 Distinguish between an optimum yield and a maximum yield.

FERTILISERS

Light and carbon dioxide are needed for plants to make carbohydrates during photosynthesis. To make other compounds, such as proteins and chlorophyll, plants need inorganic ions. They get these from the soil in which they grow.

◆ In natural environments, these ions are recycled.

◆ In cultivated environments, these ions are lost as the crop is harvested and taken elsewhere for sale. Growers can use one of two types of fertiliser to replace these lost ions.

– Inorganic fertilisers are powders produced in factories.

– Organic fertilisers are waste materials from farms, such as manure – a mixture of animal faeces, urine and bedding straw.

Table 12.3 compares the relative advantages of these two types of fertiliser.

Inorganic fertilisers (powders in solution)	Organic fertilisers (e.g. farmyard manure)
Dry, clean and without smell	Often wet, contain decomposing organic material and often have strong smell.
Concentrated sources of inorganic ions, so can be transported and applied in smaller amounts	Contain bulky materials so must be transported and applied in relatively large amounts.
Powders do not improve soil structure	Add organic material to soil, which improves soil structure as well as its water-holding capacity.
Soluble, so nutrients are immediately available to plants	Need to be broken down to release inorganic ions, so nutrients are less quickly available to plants.
When dissolved in soil water, surplus nutrients are washed into streams, ponds and lakes (leaching)	Nutrients are released more slowly, so leaching is less of a problem. Nutrients are available to plants for longer.
Costly to produce	Cheap, since farmyard waste is already available.

TABLE 12.3 A comparison of inorganic fertilisers and organic fertilisers

 5 Give two advantages of organic fertilisers over inorganic fertilisers.

EXAMINER'S TIP

Manipulating the environment in which crops are grown involves a cost to the grower. In an examination, you will be expected to realise that the important consideration is the ratio of cost to benefit. In other words, the grower is looking for the best increase in his income compared to the cost of achieving it.

Table 12.4 shows the effect of adding different quantities of fertiliser on the yield of a crop. Notice that, as with controlling environmental conditions in a glasshouse, a point is reached at which the ion concentration is no longer a limiting factor. A farmer aims for an optimum yield rather than a maximum yield.

Graph	Description	Explanation
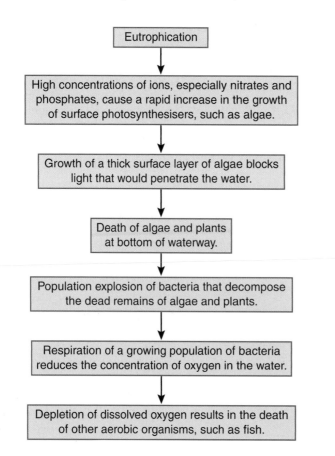 (B, C, D arrows; A arrow; Yield of crop vs Amount of fertiliser added)	◆ Between points A and B, an increase in the amount of fertiliser results in an increase in crop yield. ◆ Between points B and C, further increases in the amount of fertiliser have no effect on crop yield.	◆ Ion concentration is a limiting factor on the rate of growth. ◆ Ion concentration is no longer a limiting factor on the rate of growth.

TABLE 12.4 The effect of adding different quantities of fertiliser on the yield of a crop

6 In Table 12.4, suggest why the crop yield falls between points C and D.

7 What would you expect to happen to waterways if the use of inorganic fertilisers in the surrounding farmland was stopped?

LEACHING OF NUTRIENTS AND EUTROPHICATION

The use of inorganic fertilisers usually results in more inorganic ions being applied to the soil than crop plants can use. This leads to:

◆ leaching – dissolved in soil water, surplus ions are washed from the soil into waterways, such as streams, rivers, ponds and lakes
◆ eutrophication – the build up of inorganic ions in waterways. Figure 12.3 summarises the effect of eutrophication on waterways.

Eutrophication

↓

High concentrations of ions, especially nitrates and phosphates, cause a rapid increase in the growth of surface photosynthesisers, such as algae.

↓

Growth of a thick surface layer of algae blocks light that would penetrate the water.

↓

Death of algae and plants at bottom of waterway.

↓

Population explosion of bacteria that decompose the dead remains of algae and plants.

↓

Respiration of a growing population of bacteria reduces the concentration of oxygen in the water.

↓

Depletion of dissolved oxygen results in the death of other aerobic organisms, such as fish.

EXAMINER'S TIP

More able candidates can describe the harmful effects of fertiliser and pesticides on the environment. Weaker candidates simply state 'it harms the environment' without any further description. Make sure you show your understanding in an examination.

FIGURE 12.3 Eutrophication can have serious effects on waterways that are surrounded by agricultural land.

Controlling pests

A pest is an organism that reduces the yield of crop plants by reducing their rate of photosynthesis. Pests might do this in several ways.

◆ By competing with the crop plants. A weed is a plant that grows in a cultivated area and reduces the crop's growth rate by competing for water, inorganic ions, light or space. Competition between two species, such as a weed population and a crop population, is called interspecific competition.

◆ By eating the crop plants, so reducing the amount of photosynthetic tissue.

◆ By causing disease in the crop plants.

There are several ways in which pests can be eliminated from cultivated areas.

PESTICIDES

These are chemicals that are used to kill pest organisms. Pesticides are classed according to the type of pest they eliminate: fungicides kill fungi; herbicides kill weeds; insecticides kill insects.

Tables 12.5 and 12.6 show the properties of a successful pesticide and the different methods by which pesticides can be applied.

Property of successful pesticide	Explanation
Biodegradable	Is broken down into harmless substances in the soil and so does not accumulate in food chains (bioaccumulation)
Chemically stable	Has a long shelf life, so can be stored by farmers
Cost effective	◆ Income from sale of pesticide outweighs the cost to the manufacturer of developing and testing a new pesticide ◆ Income from increased yield outweighs the cost to the farmer of buying and applying pesticide
Specific	Kills only the target pest organism and leaves others, including humans, unharmed

TABLE 12.5 The features of a successful pesticide

Name of method	Description of method	Advantages/disadvantages of method
Contact pesticides	Sprayed directly on to the affected crop where they are absorbed by the pests that are present on the crop	◆ Relatively cheap to use ◆ Quickly washed away, so that their effects are short-lived ◆ Do not affect pests that migrate into the area after spraying. As a result, contact pesticides must be reapplied regularly.

Systemic pesticides	Sprayed directly on to the crop but do no harm to the crop plants Herbicides ◆ Absorbed by leaves and transported through bodies of weed plants ◆ Kill all tissues of weed plants, including underground parts such as roots Insecticides ◆ Absorbed by leaves and transported around bodies of the crop plants ◆ Taken in by any insect that feeds on sprayed crop plant ◆ Poison any insect that eats the sprayed crop	◆ Relatively cheap to use ◆ Specific to pest, so do not harm crop plants ◆ Relatively long-lived since they remain inside the plants ◆ Need only be applied once ◆ Affect insect pests that migrate into the area and feed on crop after spraying
Residual pesticides	Seeds of crop plant, or soil in which they will be sown, sprayed with pesticide	◆ Relatively cheap, since applied once ◆ Not washed away ◆ Remain active for a long time, killing pests before they become a nuisance

TABLE 12.6 Pesticides can be applied in several ways.

Figure 12.4 shows how long-lasting, or persistent, pesticides can have a harmful effect on the environment in which they are used. By accumulating in food chains (bioaccumulation), they kill non-target organisms. To avoid this damage, less toxic and short-lived pesticides are preferred today.

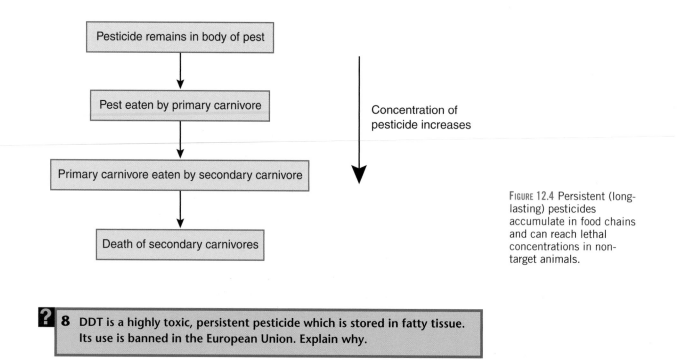

FIGURE 12.4 Persistent (long-lasting) pesticides accumulate in food chains and can reach lethal concentrations in non-target animals.

8 DDT is a highly toxic, persistent pesticide which is stored in fatty tissue. Its use is banned in the European Union. Explain why.

BIOLOGICAL CONTROL

Instead of chemical pesticides, biological control uses organisms that are natural predators or parasites of the pest.

◆ Large numbers of the biological control organism are bred or cultured and then released on to the crop.
◆ To be successful, the control organism must:
  – be specific – i.e. attack only the pest and not other, desirable organisms
  – be able to establish and maintain its population in the environment in which the crop is grown.

Although biological control programmes cause less damage to the environment than pesticide programmes, they:

◆ maintain the pest population at a low level but rarely eradicate it (see Figure 12.5)
◆ take time to show an effect.

These disadvantages are overcome by using an integrated pest control programme.

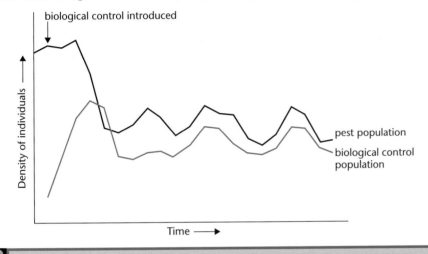

FIGURE 12.5 Biological control programmes maintain a pest population at a low level but do not eradicate the pest.

> **9** In Figure 12.5, suggest why:
> a) the populations of pest and biological control organism show regular cycles of number,
> b) the pest population is not eradicated.

INTEGRATED PEST CONTROL SYSTEM

This method is based on biological control with the occasional use of pesticides. Table 12.7 shows the relative advantages of the two aspects of an integrated system.

Advantages of using mainly biological control	Advantages of using occasional chemical control
◆ Specific to pest population ◆ Biological control population is introduced only once ◆ Pests do not become resistant to biological control population, as they would a chemical pesticide	◆ Kills pests during the period taken for the biological control population to establish itself ◆ Helps to eliminate the pest, since biological control usually maintains the pest at a low level but does not eradicate it

TABLE 12.7 Comparing the two aspects of an integrated pest control system

Figure 12.6 shows the effects of these two types of programme on a pest population.

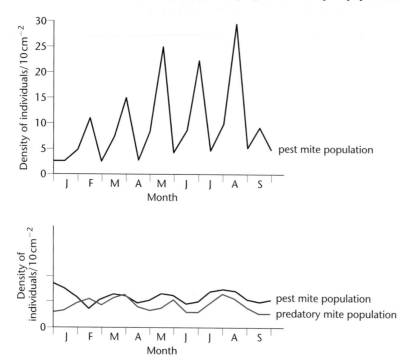

FIGURE 12.6 The effect of chemical control and biological control on a pest population

EXAMINER'S TIP

Used appropriately, pesticides have certain advantages over biological control. Many candidates seem to forget this in examinations and so cannot fully evaluate the issues involved in using the different methods. Make sure you learn the advantages *and* disadvantages of the different methods of pest control.

WORKED EXAM QUESTION

1   The diagram shows a section through the stem of a rice plant.

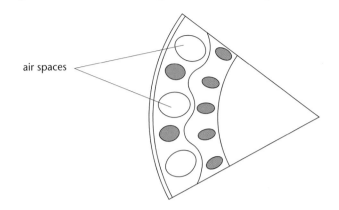

air spaces

a) Name **two** gases, present in the air spaces shown in the diagram, which can be used in chemical processes in the plant. For **each** of these gases, name **one** process in the plant which uses them.   *(4 marks)*

Gas 1   *Oxygen*

Process 1   *This gas is produced in photosynthesis by the plant.*

Gas 2   *Carbon dioxide*

Process 2   *This gas is used in photosynthesis by the plant.*

> The candidate has correctly identified both gases and the use of carbon dioxide. Unfortunately, the candidate did not give a use for oxygen, which is respiration, and so does not gain this fourth mark.

b) Describe **one** other function of the air spaces in the stem.   *(1 mark)*

Keeping the plant floating in the swamp.

> This is not an adequate answer. The candidate would have gained a mark for stating any one of the following: transports oxygen to roots/keeps leaves in light/keeps leaves in atmosphere.

c) The cells of the rice plant are more tolerant than those of other plants to high concentrations of ethanol. Explain how this tolerance is an adaptation to growth of the plant in swamp conditions.   *(3 marks)*

Swampy soils contain little oxygen so the plant's roots must respire anaerobically. Ethanol is a product of anaerobic respiration in plants. Since ethanol is toxic, the plant is at an advantage if it can tolerate high concentrations of oxygen.

> The candidate has made three valid points and gains all three marks.

(AQA 2003)

EXAMINATION QUESTIONS

1   The adult whitefly and its larvae feed on the leaves of crop plants. A parasitic wasp lays its eggs inside the whitefly larvae. When the wasp larvae hatch, they feed on the internal organs of the whitefly larvae. The wasps can be released into glasshouses to act as a biological control for the whitefly.

   a) How would the whitefly larvae reduce the yield of the crop plant?   *(2 marks)*

   b) (i) Give **two** reasons why biological control might be better than the use of chemical pesticides for controlling whitefly.   *(2 marks)*

      (ii) Give **two** reasons why the use of chemical pesticides might be better than biological control of the whitefly.   *(2 marks)*

2   The environment of crop plants can be altered to increase their growth rates and productivity.

   a) Fields which are regularly used to grow crops must have fertilisers added to them. Explain why.   *(1 mark)*

   b) The table shows a comparison of some features of organic and inorganic fertilisers.

Feature	Organic fertilisers	Inorganic fertilisers
Nutrient concentration	Low	High
Solubility	Low	High
Rate of nutrient release into soil	Slow	Rapid

   Using information from the table, explain **one** disadvantage of adding inorganic rather than organic fertiliser to farmland.   *(2 marks)*

   c) The graph shows the rate of photosynthesis of wheat plants in different conditions inside a glasshouse.

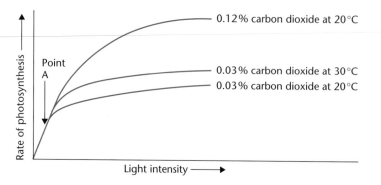

   Which factor is limiting the rate of photosynthesis at point **A**? Explain your answer.   *(2 marks)*

d) The diagram shows sections through a typical leaf of three cereal plants.

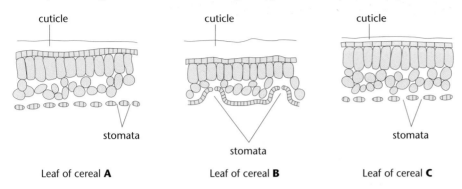

Leaf of cereal **A**   Leaf of cereal **B**   Leaf of cereal **C**

Explain which of these cereals is most likely to be able to grow in hot, dry conditions.   (*2 marks*)

(AQA 2001)

Manipulating reproduction

After revising this topic, you should be able to:

 ▶ show an understanding of the events that occur in the ovaries and endometrium during the sexual cycle of a female mammal

 ▶ describe and interpret graphs showing the roles of FSH, LH, oestrogen and progesterone in controlling the sexual cycle of a female mammal

 ▶ describe and explain the importance of the detection of oestrus in a female mammal

 ▶ explain how extracted and synthetic hormones can be used in the control of human fertility

 ▶ explain how hormones can be used in agriculture for producing embryos for transplants, for synchronising breeding and for increasing milk yield

 ▶ show an understanding of the moral and ethical issues involved in controlling reproduction.

The oestrous cycle

The reproductive system of a female mammal goes through regular cycles of events. Each cycle is called an oestrous cycle and affects two organs.

 ◆ The ovary – an organ containing connective tissue, blood vessels, hormone-producing tissue and ovarian follicles. The follicles produce the oocytes that are released at ovulation and will develop into eggs after penetration by a sperm cell.

 ◆ The uterus – a muscular organ in which one or more fetuses will implant and develop. We are mainly concerned with the lining of the uterus, the endometrium.

EXAMINER'S TIP Although the correct term for the structure which is released from an ovary at ovulation is a primary oocyte, you will not be penalised in an examination if you refer to it as an egg or an egg cell.

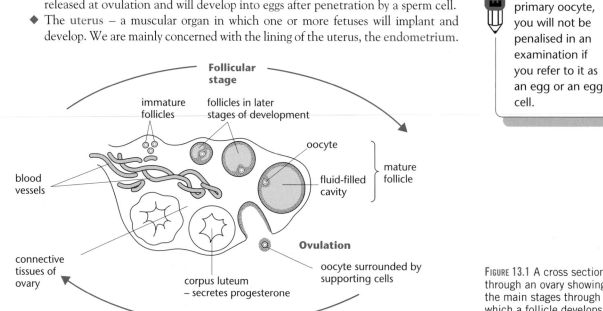

FIGURE 13.1 A cross section through an ovary showing the main stages through which a follicle develops during the oestrous cycle.

Figure 13.1 shows a cross section through an ovary. It shows the stages through which a follicle passes during one oestrous cycle. The cycle has three distinct stages: the follicular stage, ovulation and the luteal stage. Table 13.1 summarises the sequence of events which occurs during each of these stages.

Stage	Hormonal secretion	Events in ovary	Events in endometrium
Follicular stage	FSH is released and stimulates the follicle.	One or more follicles start to mature.	
	The cells in the outer region of the mature follicle secrete oestrogen.		Oestrogen stimulates an increase in thickness of the endometrium and stimulates the development of more blood vessels.
	Once it reaches a particular concentration in the blood, oestrogen inhibits further secretion of FSH and stimulates the secretion and storage of LH by the pituitary.		
Ovulation	A surge occurs in the blood concentration of LH as stored LH is released by the pituitary.	An oocyte is released from one or more mature follicles.	
		The empty follicle becomes a corpus luteum.	
Luteal Stage	Cells in the corpus luteum secrete oestrogen and progesterone.		Progesterone stimulates the endometrium to: ◆ grow thicker ◆ produce more blood vessels ◆ develop glands that secrete a nutritive fluid.
	High blood concentrations of oestrogen and progesterone inhibit further release of FSH and LH.	Without FSH and LH, the corpus luteum degenerates and stops the production of oestrogen and progesterone.	

TABLE 13.1 The events that occur during one oestrous cycle in which pregnancy does not occur. The arrows in the table show the sequence of these events.

Four hormones are named in Table 13.1. Two are released from the anterior part of the pituitary gland, located just under the brain, and two are released by the ovaries.

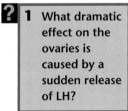

? 1 What dramatic effect on the ovaries is caused by a sudden release of LH?

1 Follicle stimulating hormone (FSH) is released by the anterior pituitary gland and stimulates the ovaries and the endometrium.

2 Luteinising hormone (LH) is produced and stored in the anterior pituitary. Its sudden release stimulates the endometrium and has a dramatic effect on the ovaries.

3 Oestrogen is released from mature ovarian follicles and by corpora lutea (the plural of corpus luteum). It affects the endometrium and the anterior pituitary gland.

4 Progesterone is released by corpora lutea and affects the endometrium and the anterior pituitary gland.

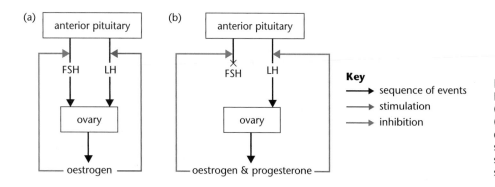

Key
→ sequence of events
→ stimulation
→ inhibition

Figure 13.2 The interaction between hormones during (a) the follicular phase and (b) the luteal phase of the oestrous cycle. Different styles of arrow represent a sequence of events, stimulation and inhibition.

The interaction between these hormones is complex. You must understand this interaction and be able to interpret data relating to it. Figure 13.2 shows one way in which the interaction of the hormones controlling the oestrous cycle can be represented. These two diagrams represent the stimulation and inhibition of hormones during the follicular stage and the luteal stage of the cycle. Figure 13.3 is another common way of representing the hormones that regulate the oestrous cycle.

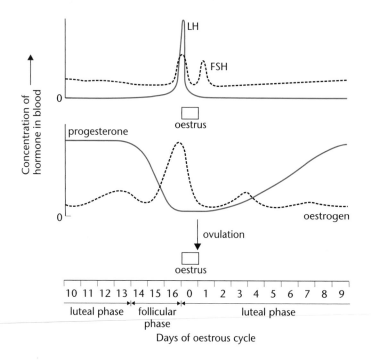

Figure 13.3 The hormonal events that control the oestrous cycle in a female mammal

EXAMINER'S TIP

Rather than learn Figure 13.3, make sure you understand it. For example, make sure you can use:
♦ the FSH curve to explain the rise in the oestrogen curve prior to ovulation
♦ the oestrogen curve to explain the surge in the LH curve
♦ the LH curve to explain the shape of the progesterone curve after ovulation
♦ the curves for oestrogen and progesterone to explain the shapes of the FSH and LH curves after ovulation.

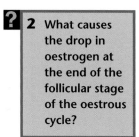

2 What causes the drop in oestrogen at the end of the follicular stage of the oestrous cycle?

Oestrus

As well as physiological changes, the behaviour of female mammals changes around the time of ovulation (see Table 13.2).

◆ These changes are called oestrus and are brought about mainly by oestrogen. More commonly, a female is said to be 'on heat' during oestrus.

◆ You need to be able to describe how these changes can be identified by a farmer and used to detect when an animal is in oestrus. You need to be able to do this only for one, named animal.

◆ You also need to realise that recognising oestrus is helpful to a farmer because it signals the time at which animals should be inseminated, either through mating or by artificial insemination. Since a mature female gamete lives for only a short period of time, insemination is likely to result in successful fertilisation only if it takes place within hours of ovulation.

> **? 3** Suggest why fertilisation can occur only for a few hours after ovulation.

Physiological changes	◆ Dilation of neck of uterus (the cervix). ◆ Change in texture of mucus produced by cervix so that it becomes thinner and more slimy. As a result, it flows from the cervix through the entrance of the vagina and can be seen by farmers.
Behavioural changes	◆ Increased restlessness. The cow moves about more and feeds less. In the early stages, she might mount other cattle from behind. ◆ Decreased aggression towards other members of the herd. In the later stages of oestrus, the cow will remain still and allow other cattle to mount her.

TABLE 13.2 Some of the changes that occur in cattle at the time of oestrus

> **? 4** What advantage is gained by:
> a) a bull in recognising oestrus among the cows in a wild population of cattle, such as wildebeest;
> b) a farmer in recognising oestrus in a herd of domesticated cows?

Controlling human fertility

Hormones, and synthetic hormones, can be used to control human fertility in two ways:

◆ as contraceptives – preventing unwanted pregnancies
◆ to treat infertility – helping couples who are having difficulty conceiving a child.

CONTRACEPTIVES

High blood concentrations of oestrogen and progesterone inhibit the release of FSH from the anterior pituitary. Table 13.3 shows information about three types of contraceptive pill that contain hormones.

Type of pill	Contents of pill	Effects of pill
Combined contraceptive pill – taken daily	Mixture of oestrogen and progesterone	◆ Increases a woman's blood concentration of oestrogen and progesterone. ◆ These hormones inhibit FSH release. ◆ No follicles will mature in a woman's ovaries and she will not ovulate. ◆ High oestrogen concentrations have been linked to a number of harmful side effects amongst women taking the combined contraceptive pill, including an increased risk of thrombosis and strokes.
Mini pill – taken daily	Progesterone	◆ Stops production of oocytes. ◆ Interferes with the process of meiosis – alone it does not inhibit FSH release.
Morning after pill – effective if taken up to 72 hours after intercourse	High concentration of progesterone	Stops implantation occurring

TABLE 13.3 These three types of contraceptive pill contain female hormones.

> **?** **5** **Explain why taking the mini pill carries less of a risk to a woman than taking the combined contraceptive pill.**

Not all contraceptives are taken in pill form. Some can be inserted into the fatty tissue beneath a woman's skin, where they slowly release their hormones into her bloodstream.

> **?** **6** **Suggest why an implanted contraceptive might be more effective than an oral contraceptive.**

TREATING INFERTILITY

Infertile couples have trouble conceiving children despite regular sexual intercourse. There are many causes of infertility, some relating to the woman (e.g. failure to ovulate) and some relating to the man (e.g. failure to produce sufficient quantities of viable sperm).

If a normally healthy woman is failing to ovulate, she can be treated in two ways.

♦ With FSH. This causes her to ovulate. The technique is often used to collect several oocytes from a woman prior to fertilisation by sperm in laboratory glassware (*in vitro* fertilisation or IVF). A fertilised oocyte is then replaced into the woman's uterus.

♦ With a drug, such as clomiphine, which prevents the inhibition of FSH release by oestrogen.

? **7 How would clomiphine help a woman to conceive?**

Controlling reproduction of domestic animals

The productivity of domestic animals is the rate at which they produce more muscle, produce eggs or release milk. Increasing productivity produces more food for humans and increases the profits made by farmers. You need to be familiar with three ways in which productivity can be increased by controlling the reproduction of domestic animals.

1 TRANSPLANTING EMBRYOS

This technique is used to enable large numbers of females to become pregnant with the embryos of a single mother that carries a desirable genetic characteristic. Figure 13.4 summarises this technique.

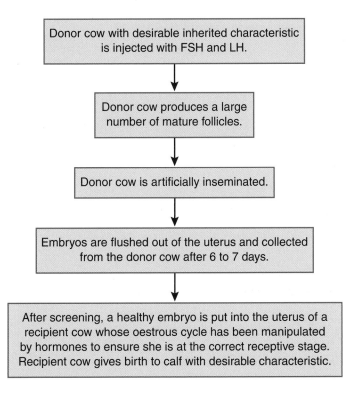

FIGURE 13.4 Embryo transplantation. A cow's oestrous cycle is 21 days, during which she normally produces only one egg. A cow produces only two or three calves in her lifetime. Embryo transplantation speeds up the rate at which a cow can be used to improve the genetic status of a herd.

? **8 What is the advantage of transplanting embryos from a donor cow into a recipient cow rather than allowing the recipient to have calves formed by the fertilisation of her own eggs?**

2 SYNCHRONISING BREEDING

Farmers pay fees to a registered vet who will artificially inseminate a group of female livestock. Farmers also attend their animals as they give birth. It is cost effective and time effective to synchronise the breeding of animals so that they are inseminated at the same time and give birth around the same time.

Figure 13.5 shows how progesterone can be used to synchronise ovulation in sheep.

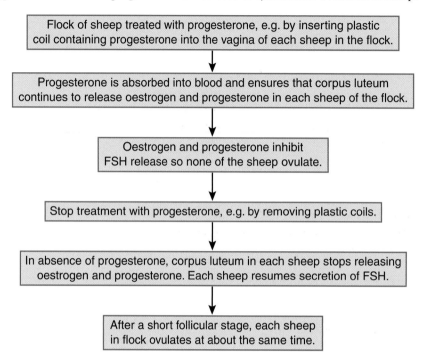

FIGURE 13.5 Synchronisation of breeding in sheep. The use of progesterone ensures that all sheep in the flock are ready for insemination at the same time and give birth to lambs at about the same time.

3 INCREASING MILK YIELD IN COWS

A genetically engineered growth hormone, bovine somatotrophin (BST), stimulates:
- growth of a cow's udders
- more of the protein, carbohydrate and lipid from a cow's diet to be used for milk production.

Injected into their bloodstream, BST enables cows to produce more milk over a longer period of time than untreated cows.

Moral and ethical considerations

Whatever your own opinions, make sure you understand the moral and ethical issues concerning the use of biotechnology to manipulate reproduction. These mainly concern the control of human fertility and include the following:
- Religious objections to the use of contraception. For example, Catholicism, a major Christian faith, forbids its followers to use contraception.
- Ethical objections against the use of human embryos, produced by IVF, in subsequent research programmes. Many people strongly believe that the life of a human embryo is as sacrosanct as that of a child or an adult. Most countries have laws restricting the use of human embryos produced by IVF.
- Concerns that limited funding for health programmes is diverted from life-threatening use to help infertile couples.

WORKED EXAM QUESTION

1   The graph shows how the diameters of the follicles and corpora lutea vary during the oestrous cycle of a pig.

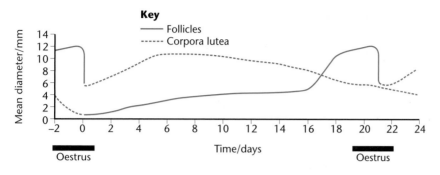

a) In this cycle, ovulation occurred on day 21. Explain how the graph gives evidence for this.   *(2 marks)*

The follicle gets smaller when it loses an egg at ovulation. The graph shows that the mean diameter of the follicles drops suddenly on day 21 so they must have lost their eggs then.

> The candidate has selected the correct information and gains both marks.

The table shows how the concentration of the hormones progesterone and oestrogen varies during the oestrous cycle of the pig.

Time/ days	Concentration of hormone in the blood/ arbitrary units	
	Progesterone	Oestrogen
0	1.8	1.0
2	4.7	0.9
4	9.0	0.9
6	12.1	0.9
8	12.0	0.9
10	8.7	0.9
12	3.5	0.9
14	0.9	1.7
16	0.6	2.1
18	0.6	6.0
20	1.1	3.2
22	3.7	0.9
24	7.5	0.9

b) Use the table and the graph to explain why the presence of oestrogen in the female pig's urine can be used to predict the start of oestrus.   (*2 marks*)

The table shows that the concentration of oestrogen reaches its highest value on day 18. This is one day before oestrus starts on the graph.

Again, the candidate has used the stimulus material and selected the correct information.

c) Describe and explain the relationship between the diameter of the corpora lutea and the concentration of progesterone in the pig's blood between days 0 and 14.   (*2 marks*)

The concentration of progesterone in the table rises between days 0 and 6. The graph shows that the diameter of the corpora lutea increases at the same time. After day 6 the concentration of progesterone starts to fall up to day 14 and the diameter of the corpora lutea falls slightly during this time.

This is an excellent description of the relationship and gains one mark. The candidate has not given an explanation for the relationship and so does not get the second mark. A suitable explanation would be: corpus luteum produces progesterone. Notice that, so far, few marks have been awarded for recall. Most have been for data interpretation. You need to practise this skill using graphs and tables.

d) Use the data and your own knowledge to explain how changing concentrations of oestrogen and progesterone regulate the oestrous cycle.   (*6 marks*)

FSH stimulates follicles to develop in the ovaries. The ovaries secrete oestrogen and this inhibits FSH but stimulates LH. LH causes ovulation.

A question like this gives examiners an opportunity to test recall and understanding in an extended free-response answer. It is worth practising answers to questions like this because they are likely to recur. This candidate has gained 4 of the 6 available marks in three short sentences. There is often no need to write at length if you are clear about the answer.

This candidate's answer is incomplete and does not contain enough information to gain full marks. Also, one part of the answer is too vague to gain a mark – it is the follicles, rather than the ovaries, which release oestrogen.

Further marks could be obtained for: LH stimulates release of progesterone; progesterone inhibits FSH/LH; FSH increases when progesterone falls/on day 16.

(*AQA 2003*)

EXAMINATION QUESTION

1 a) The oestrous cycle in a female mammal is controlled by hormones. Describe the part played by FSH and LH in the control of the oestrous cycle.

(5 marks)

b) The oestrous cycle of female sheep can be synchronised by giving them low doses of progesterone. When the treatment is stopped, the sheep come into oestrus a short time later.

(i) Explain why low doses of progesterone prevent oestrus in sheep.

(2 marks)

(ii) Explain why sheep come into oestrus a short time after progesterone treatment is stopped.

(2 marks)

c) The effect of progesterone treatment on the time when lambs were born was investigated. The graph shows the dates of lambing in a group of sheep in which the oestrous cycles were synchronised and in a control group.

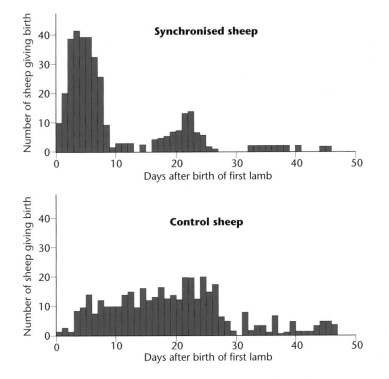

Using information in the graph:

(i) suggest the advantage to a farmer of synchronising oestrous cycles in sheep;

(2 marks)

(ii) estimate the length of an oestrous cycle in sheep. Explain how you arrived at your answer.

(2 marks)

d) Describe how the control group of sheep should have been treated. (2 marks)

(AQA 2002)

ANSWERS TO IN-TEXT QUESTIONS

CHAPTER 1

1 Prokaryotic cells do not have a nucleus whereas eukaryotic cells do. The genetic material of prokaryotic cells is not arranged in chromosomes whereas it is in eukaryotic cells. Prokaryotic cells lack the membrane-bound organelles/lack a named example of a membrane-bound organelle found in eukaryotic cells.

> **EXAMINER'S TIP**
>
> When asked to make a comparison or describe a difference between two things, remember to mention both structures being compared. In answer 1 above, you needed to write something about the prokaryotic cell and something about the eukaryotic cell.

2 The ability of a microscope to distinguish between structures that are close together.

3 Any two of the following answers would be adequate. The processes involved in preparing specimens might distort their true structure. Placing specimens in a vacuum might distort their structure. Bombarding specimens with electrons might distort their structure. Specimens are always dead and so we could never check the structure of a living specimen using a transmission electron microscope.

4 Magnification tells us only apparent size ÷ actual size. The estimated size should be a value with units.

5 0.5 mm (2 mm ÷ 4)

6 Actual length $= \dfrac{\text{Measured length}}{\text{Magnification}}$

$= \dfrac{0.048 \text{ m}}{5400}$

$= 8.9 \; \mu\text{m}$

7 Proteins made by the ribosomes are transported to the Golgi apparatus, where they are stored and modified prior to secretion. The activities of the ribosomes and of the Golgi apparatus use ATP, which is made by the mitochondria.

8 These are the regions that have allowed electrons to pass through the specimen. The dark regions have not let electrons pass through them.

9 a) To slow down degeneration of the organelles by the action of enzymes.

b) To prevent distortion of the organelles following water entry or loss by osmosis.

10 a) Mitochondria or lysosomes

b) To remove the unwanted nuclei and chloroplasts that would contaminate the pellet she wanted.

CHAPTER 2

1 A metabolite is a substance that is needed for one of the cell's reactions (metabolism).

2 a) The membrane is too small to be resolved by any existing microscopes. Therefore its structure must be deduced from other, indirect, evidence.

b) It is fluid because the molecules move about within it. It is a mosaic because it consists of a patchwork of different types of molecules (e.g. phospholipids, proteins and glycoproteins).

3 Receptors (e.g. for a hormone), carriers (in facilitated diffusion), channels, pumps (e.g. sodium-potassium pumps)

4 a) Liver cells have insulin receptors on their plasma membranes; cells in the heart do not.

b) Description: Insulin increases the rate of uptake of glucose into the cells.
Explanation: More carrier proteins join the plasma membrane (from vesicles in the cytoplasm). This increases the number of channels through which glucose can enter the cell by facilitated diffusion.

5 Facilitated diffusion involves protein carrier molecules in the membrane whereas diffusion does not.

6 a) Solution B

b) Solution B (Note that a higher water potential means a less negative value.)

7 B ⇒ A. Osmosis occurs from a solution with a high (less negative) water potential to one with a low (more negative) water potential.

8 a) Into the cell. Osmosis occurs from a solution with a high (less negative) water potential to one with a low (more negative) water potential.

b) (i) Since they prevent cell wall formation, newly formed bacterial cells will not be able to withstand an intake of water by osmosis and will burst.

(ii) Human cells do not produce the chemical that is found in bacterial cells and so the antibiotic does not disrupt their metabolism.

9 Active transport uses energy from the breakdown of ATP, facilitated diffusion does not. Active transport occurs up a concentration gradient, facilitated diffusion occurs down a concentration gradient.
10 a) Yes. It had been made in the cytoplasm and must have passed through the plasma membrane into the vesicle.
   b) No. Since it is still inside the vesicle, it has not yet crossed the membrane into the cytoplasm.

CHAPTER 3

1 Two molecules of glucose join together in a condensation reaction, so that one molecule of water (H_2O) is lost.
2 Because one molecule of water is lost from the glucose molecules.
3 a) glycosidic bond, b) peptide bond
4 The nature of the R group
5 Heat breaks the hydrogen bonds and disulphide bridges that hold the enzyme's shape in place. As a result it loses its properties.
6 A triglyceride is not a polymer since it is not a chain made from repeated sub-units (monomers).
7 a) Boil a fresh sample of sucrose solution with dilute hydrochloric acid. Cool and add enough sodium hydrogencarbonate to neutralise the acid. Then add Benedict's solution and boil briefly in a water bath.
   b) The sucrose is hydrolysed to its constituent monosaccharides, in this case glucose and fructose. Unlike sucrose, these molecules are reducing sugars.
8 The pigments in the ink would also be separated by the chromatography and this would confuse the result. A pencil mark will be unaffected.
9 Measuring to front edge of each spot,

$$Rf = \frac{32\,mm}{36\,mm} = 0.89$$

CHAPTER 4

1 Enzyme action relies on the shape of the active site. Each active site must match (be complementary to) a different substrate and so each enzyme must be a different shape. Only protein can create enough different shaped molecules to allow this to happen.
2 An enzyme brings reacting molecules together so that less energy is needed to break or form bonds.

3 If the enzyme was a rigid molecule, when a non-competitive inhibitor combined with it, there would be no change of shape. Therefore, the active site would stay the same shape and the enzyme would still function.
4 Enzymes are not altered by the reactions they influence and so are reusable. Also they act very quickly to break down or build up molecules.

CHAPTER 5

1 glucose + adenosine diphosphate + phosphate + oxygen → adenosine triphosphate + carbon dioxide + water
2 pharynx, epiglottis, trachea, bronchus, bronchiole, alveolus
3 a) Mucus and attached substances build up in the lungs, causing irritation.
   b) Mucus traps bacteria and viruses and without effective cilia, these gain entrance to the lungs.
4 a) Rate of diffusion proportional to (surface area × difference in concentration)/thickness of gas exchange surface.
   b) Many alveoli increase gas exchange surface area; blood bringing CO_2 to, and taking O_2 from, the alveoli maintains a concentration gradient; alveoli have a single layer of squamous cells giving a thin exchange surface.
5 Carbon dioxide is in higher concentration in the blood plasma than in the air in the alveoli. It diffuses down its concentration gradient into the air in the alveoli.
6 a) The lining of the trachea, bronchi and bronchioles are not permeable, so no gas exchange occurs in this part of the respiratory system. This part of the respiratory system is often called 'dead space' and has a volume of about $150\,cm^3$.
   b) The alveoli are permeable to water, which evaporates and diffuses into alveolar air.
7 Contraction of the internal intercostal muscles would pull the ribs downwards and inwards, increasing the pressure on the lungs and causing a large volume of air to be exhaled.
8 The lungs deflate during expiration. As a result, the stretch receptors in the lungs are no longer stimulated and, therefore, no longer inhibit the inspiratory centre.
9 $6000\,cm^3\,minute^{-1}$ ($500\,cm^3 \times 12$ breaths $minute^{-1}$), which you could write as $6\,dm^3\,minute^{-1}$

10 a) The rate of respiration increases in exercising muscle cells. As CO_2 is a waste product of this respiration, more will be produced during exercise.

b) CO_2 dissolves in water in the plasma to form carbonic acid, which dissociates to produce hydrogen ions (H^+) and hydrogencarbonate ions (HCO_3^-). An increase in the concentration of H^+ ions reduces the pH value.

CHAPTER 6

1 Artery – because the arteries receive blood directly from the heart.

2 The capillary wall is one cell thick; it consists only of the epithelium.

3 a) Pressure in the capillary is still too high to allow a net movement of fluid back into the blood system. Tissue fluid builds up resulting in a condition called oedema.

b) The water potential of the blood is too high and it contributes to the net movement of fluid out of the capillary. Once again fluid is retained in the tissues.

4 Plasma contains soluble organic molecules, plasma proteins, red and white blood cells.
Tissue fluid does not contain plasma proteins or red blood cells.
Lymph is similar to tissue fluid but it will contain higher numbers of white blood cells – lymphocytes.

5 Any two from: endothelium; muscle; elastic connective tissue.

6 Right atrium – right ventricle – plumonary artery – lungs – pulmonary vein – left atrium – left ventricle

7 a) Valves closing
b) 0.15/0.17 seconds
c) 80 beats per minute

CHAPTER 7

1 The shape of their active site is complementary to the shape of their substrate.

2 Many reactions release heat, which would increase the temperature inside the fermenter. An increase in temperature would denature the enzymes in the fermenter.

3 a) To make organic compounds, which are all based on carbon.

b) To make nitrogen-containing organic compounds, such as amino acids.

4 Conditions in which only the desired microorganism can be introduced into a container and no microorganisms can leave a container.

5 Downstream processing for intracellular enzymes is more complex because the microbial cells must be ground up to release their enzymes. The enzyme must then be separated from the sub-cellular fragments and from other compounds in the mixture.

6 Industrial reactions often occur at high temperatures. Thermostable enzymes are resistant to these high temperatures and to changes in temperature caused by the release of heat during the catalysed reaction.

CHAPTER 8

1 Protein – amino acids; carbohydrates – monosaccharides

2 Condensation is the joining of two smaller molecules to make a larger molecule with the removal of a water molecule.

3 Both are made up of nucleotide monomers, which contain a phosphate, a nitrogenous base and a pentose sugar.

4 DNA – maintains the genetic code for the lifetime of the cell and is passed on to the next generation (i.e. it is the inherited material).
RNA – controls the production of a protein; only a little of each protein is needed at once.

5 ◆ RNA has ribose sugar instead of deoxyribose sugar and uracil instead of thymine, so it is impossible to make one from the other.
◆ Also DNA is only used as a template and remains intact at the end of the process.

6 Hydrogen; covalent; ionic; disulphide

7 Replication is when a DNA molecule is copied to make two molecules of DNA.
Transcription is when DNA is used as a template to make a molecule of mRNA.

8 Nitrogenous base

9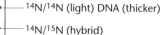

$^{14}N/^{14}N$ (light) DNA (thicker)
$^{14}N/^{15}N$ (hybrid) DNA (no change)

CHAPTER 9

1. a) Clone – a group of genetically identical cells produced by mitosis of a single parent cell.
   b) Homologous pair – the two copies of every chromosome in a diploid cell. One copy is derived from the gametes of each parent.
   c) Haploid – a cell which has only one copy of each pair of homologous chromosomes.
2. a) 46 – the cells are clones produced by mitosis.
   b) 23 – the gamete is formed by meiosis.
3. Its chromosomes would not be clearly visible/it would not have a spindle/the nuclear envelope would still be intact.
4. D, A, C, B.
5. Mitosis – it has the narrowest bandwidth.
6. During spore formation – the spores have half the number of chromosomes of the parent cell.

CHAPTER 10

1. The diagram shows only the primary structure (sequence of amino acids).
2. Insulin-producing cells (in the islets of Langerhans in the pancreas) will have large amounts of mRNA for insulin in their cytoplasm. It might be easier to collect this than to find the insulin gene among the 46 chromosomes in a human cell.
3. a) Reverse transcriptase catalyses the formation of DNA from RNA.
   b) DNA polymerase catalyses the condensation of DNA nucleotides during DNA formation.
   c) Restriction enzymes hydrolyse DNA at specific recognition sequences.
4. AACT (you could have used the other DNA strand and correctly replied TTGA).
5. A small, circular piece of DNA that is found in some bacteria.
6. A transgenic organism contains nucleic acid from another organism.
7. a) Since they grow on the medium, they must contain the gene for resistance to antibiotic A, which is found in the plasmid.
   b) Since they do not grow in the medium, they do not have a functional copy of the gene for resistance to antibiotic B. This must mean that this gene in this plasmid has been broken by the insertion of the human insulin gene.

CHAPTER 11

1. Any two from: acid of stomach; enzyme in saliva; cilia and mucus in lungs.

2. Antibodies function because their shape complements the antigen. Only proteins can form a sufficiently large range of shapes to match the range of antigens.
3. Vaccination introduces antigens, which will stimulate a humoral response. This will result in the storage of memory B-cells so when the antigen is presented again (on the surface of a pathogen), a secondary response rather than a primary response occurs.
4. a) Group O has no antigens on its red blood cells and will not stimulate an immune response.
   b) Group AB has no antibodies in the plasma and so does not respond to any antigens in donated blood.
5. Red blood cells have no nucleus and therefore no DNA.
6. Yes; there are a number of bands that match the father's bands; importantly these have not come from the mother.

CHAPTER 12

1. Water evaporates from the cells inside a leaf and diffuses out into the surrounding air through the stomata. A small number of stomata reduces the area through which water vapour can diffuse. The air around sunken stomata becomes saturated with water vapour, reducing the diffusion gradient out of the plant.
2. Any environmental resource that is needed for photosynthesis, and that is in short supply, is a limiting factor. Its low value limits the rate of photosynthesis.
3. In both, light intensity is a limiting factor over the initial, steep parts of the curves. Where both curves are level, the higher rate of photosynthesis in curve E is related to the higher carbon dioxide concentration, since they now have the same light intensity and temperature. Carbon dioxide concentration is the limiting factor.
4. An optimum yield produces the best income in relation to additional expenditure on maintaining conditions in the glasshouse. A maximum yield produces the highest yield possible, regardless of additional expenditure on maintaining conditions in the glasshouse. A grower might actually lose money trying to achieve a maximum yield.
5. Any two from: improve soil structure; improve water-holding capacity of soil; nutrients available to plants for longer; cheap on farms where farmyard waste is already available.

6   High concentrations of ions make the water potential of the soil more negative so plants lose water through their roots by osmosis. Alternatively, high ion concentrations might be directly toxic to the plant.

7   The growth of surface photosynthesisers would slow. More light would penetrate to the bottom of the waterway, enabling other plants to grow. The rate of bacterial decomposition would decrease, making more oxygen available for the respiration of, for example, fish.

8   Fatty tissue is not broken down. DDT accumulates in the fat of predators, reaching lethal levels.

9   a)   When numbers of pests are low, there is less food for the biological control organisms, leading to their death. When the numbers of biological control organisms are low, fewer pests are killed.

    b)   Low population numbers of the pest lead to lower population numbers of biological control organisms, reducing their efficiency in killing pests.

CHAPTER 13

1   Ovulation

2   The fall in the concentration of FSH. Oestrogen release is stimulated by a high concentration of FSH but a high concentration of oestrogen inhibits FSH release.

3   Unless fertilised, the oocyte dies.

4   a)   The male recognises which females are ready to mate, e.g. by the smell of the mucus released from the vagina. This increases the likelihood of successful fertilisation and means that the male does not waste energy by attempting to mate with infertile females.

    b)   The farmer does not waste money on vet's fees by inseminating infertile cows.

5   The mini pill contains no oestrogen. It is the oestrogen in the combined contraceptive pill that seems to be linked to increased risk of thrombosis and strokes.

6   To be effective, the contraceptive must maintain a blood concentration of hormones above a critical value to prevent conception. An implanted contraceptive releases its hormone over a long period of time and so maintains this critical concentration of hormone. With a daily pill, the blood hormone concentration falls below the critical value after about 27 hours. This means that a woman runs a risk of conception if she fails to take the pill before her blood hormone level falls below the critical value. Apart from forgetting to take an oral contraceptive, a woman might lose the hormones in the pill she has recently swallowed if she vomits.

7   By preventing the inhibition of FSH, a woman will secrete this hormone so that follicles will develop in her ovaries.

8   The donor cow carries a desired, inherited characteristic but the recipient cows do not. Embryos from the donor cow are likely to carry the desired characteristic of the donor cow. High milk yield is an example of a desired characteristic.

ANSWERS TO EXAMINATION QUESTIONS

CHAPTER 1

1 a) **A** Carries the (genetic) code/genetic instructions/DNA/makes mRNA/transcription/makes ribosomes.
       B Links amino acids/synthesises/makes protein.
       C Involved in modifying/packaging protein/forms glycoproteins/forms vesicles 3
   b) (i) Mitochondrion   1
           0.01% as opposed to 0.003%   1
           (any valid approach will be accepted, but it must be clear as to what the calculations relate)
       (ii) With electron microscopes sections must be cut;
           Cisternae are joined to each other;
           Outside plane of section.   2 max
       (iii) Protein synthesis requires energy/ATP;   1
           Mitochondria release energy/make ATP;   1
           From respiration.   1
           (no credit would be awarded for second point if answer refers to mitochondria making/producing energy)
                              Total marks = 10

CHAPTER 2

2 a) (Crush in) ethanol/alcohol;
       Add (to) water *(order of adding is critical for this point)*;
       Emulsion/white colour.   3
   b) (i) Glycerol/glyceride   1
       (ii) Phospholipid has phosphate/phospholipid only has two fatty acids.   1
       (iii) Phosphorus/P   1
   c) (i) <u>Both</u> membranes contain phospholipid/lipid (bilayer).   1
       (ii) Glucose unable to pass through artificial membrane as not lipid soluble;
           Glucose transported by proteins;
           (Proteins) found in plasma membrane/not found in artificial membrane.   2 max
                              Total marks = 9

CHAPTER 3

1 a) (i) NH_2 1
       (ii) Two peptide bonds/reference to specific feature such as C=O/R groups appearing three times.   1

   b) (i) Line shown as clearly below origin.   1
       (ii) Only two different amino acids/two amino acids are the same.   1
       (iii) Rf = $\dfrac{\text{distance moved by A/substance}}{\text{distance moved by solvent front}}$ 1
       (iv) Distance moved by spot variable;
           Depends on time/distance moved by solvent;
           Rf is a ratio;
           and will be constant/same in different chromatograms.   2 max
                              Total marks = 7

CHAPTER 4

1 a) (i) Divide amount of product produced by time taken/calculate gradient/slope of graph.   1
           (no mark will be awarded for a numerical answer without supporting calculation)
       (ii) Higher temperatures mean molecules have more (kinetic) energy;
           (the idea of molecules is required for a mark)
           Move faster;
           Greater chance of collision (between enzyme and substrate);
           More chance of enzyme-substrate complex being formed.   3 max
   b) At 65°C enzyme has been denatured; description of denaturing.   1
   c) To maintain a constant pH.   1
                              Total marks = 6

CHAPTER 5

1 a) (i) Tidal volume.   1
       (ii) Multiply **A**/tidal volume/volume of breath by number of breaths per minute/breathing rate.   1
           (error in a) (i) will be penalised once only)
   b) (i) Sends <u>more</u> impulses to …;
           Diaphragm/intercostal muscles;
           Increases rate of inspiration/causes more frequent contraction.
       (ii) Sends impulses to …;
           Sinoatrial node/SAN/pacemaker;
           Increases rate of discharge/heart rate. 4 max
   c) (i) Diffusion   1
       (ii) Not normally present/needed   1
           Any detected must have come from this test.   1
   d) Longer diffusion pathway/takes longer to diffuse/slower rate of diffusion.   1
                              Total marks = 10

CHAPTER 6

1 a) (i) Pressure in right ventricle is higher than
           in right atrium.   1
       (ii) Valve drawn closed   1
    b) The impulse slows at the AV node;
        to allow the complete contraction of the
        atrium.   2
    c) No stimulus from the cardiac nerve/
        sympathetic NS;
        stimulus from the vagus nerve/
        parasympathetic NS;
        to the SAN/sino atrial node.   3

                          Total marks = 7

CHAPTER 7

1 a) (i) Optimum growth of culture;
           Prevents denaturation of enzymes.   2
       (ii) (Removes) heat <u>produced</u> by
            microorganisms (during respiration)/
            in reactions;
            Prevents denaturation of enzymes.   2
    a) Other microorganisms may compete with
        desired microorganism (for nutrients);
        Reduces yield of product.
        or
        Other microorganisms may produce by-
        products/alter pH;
        Which will contaminate the product/
        dilute product/denature enzymes.
        or
        Other microorganisms may be pathogenic
        to culture;
        Fewer microorganisms producing enzymes.
        or
        May produce toxins;
        Fewer microorganisms producing
        enzymes.   2 max
        (If candidates refer to the enzyme as a
        microorganism, do not award credit.
        Can score 2 reasons or 1 reason plus
        amplification.)
    b) Downstream processing is more complex;
        Need to break open cells;
        Extract enzyme from other substances
        present.   2 max

                          Total marks = 8

CHAPTER 8

    a) (i) P – Adenine
             Q – Guanine
             R – Thymine
             S – Cytosine   2
        (ii) Bases are paired.   1
        (iii) Only a part of the DNA strand is
              transcribed/ mRNA represents the
              transcription of one gene, DNA is
              made up of many genes.   1
    b) 608   1
    c) Makes the DNA more stable/during replication
        allows two identical copies to be made.   1

                          Total marks = 6

CHAPTER 9

1 a) (i) 20   1
       (ii) 10   1
       (iii) 10   1
    b) (i) (Daughter) chromatids will not
             separate/centromere won't divide;
             <u>Centromere</u> attaches to spindle fibres 2
             (an incorrect answer is 'chromosomes
             can't be pulled apart'. References to stages
             of mitosis will be ignored.)
        (ii) Red blood cells formed/produced by
              mitosis.   1

                          Total marks = 6

CHAPTER 10

1 a) (i) Difficulty of finding one gene among all
             the genes in the nucleus/large amounts
             of mRNA coding for insulin will be
             present in insulin producing cells/idea
             that mRNA will be 'edited'.   1
        (ii) Reverse transcriptase   1
        (iii) AGTTGG   1
    b) Joins the gene for insulin into the plasmid.   1
    c) Allows transformed bacteria to be separated
        from non-transformed;
        Further detail e.g. transformed bacteria
        survive when antibiotic applied to medium.   2

                          Total marks = 6

CHAPTER 11

a) Hydrogen bonds   1

b) (i) DNA polymerase will not bind to a single strand.   1

    (ii) TATCCGTC   1

c) 64   1

d) DNA will be present in contaminating cells; PCR will duplicate any DNA.   2

Total marks = 6

CHAPTER 12

1 a) (Damaged leaves result in) reduced photosynthesis; (feeds on/removes) organic materials/named example needed for plant (growth).   2

b) (i) Wasp can find white fly/pesticide might not reach white fly larva; wasp is specific/pesticide kills or harms other species; pesticide can contaminate human food/human during applications; pest might develop resistance to pesticide; pesticide must be re-applied.   2 max

    (ii) Chemical kills all pests/predator dies out if all pests killed; pesticide is resistant/biological control takes longer; pesticide can be applied locally/on specific plants/biological control cannot be localised on field crops.   2

2 a) Cropping/harvesting removes nutrients.   1

b) Easily lost/leached/washed out; so need extra application/causes eutrophication /good description of eutrophication.

or

More negative/lower WP/solute potential (in soil); Toxic/osmotic effect.   2

c) Light intensity; Because increasing light (intensity) causes increasing rate.   2

d) Leaf B – no mark Deep/thick cuticle/small number of/sunken stomatal/pits. Slows down/reduction in water loss/transpiration.   2 *(references to leaf rolling/motor cells will be ignored)*

Total marks = 13

CHAPTER 13

1 a) FSH stimulates growth of a follicle; Developing follicle produces oestrogen; (FSH) and LH bring about ovulation/oestrus; LH stimulates formation of corpus luteum; LH stimulates production of progesterone; Fall in LH/FSH means oestrogen production no longer stimulated.   5 max

b) (i) Progesterone inhibits FSH;   1 No follicles develop.   1

    (ii) Causes rise in FSH/inhibition of FSH removed;   1 Stimulates follicle development.   1

c) (i) (Ewes) produce lambs at similar times/inseminate at same time; Allows farmer to prepare for many births at same time/employing extra labour/can give all ewes similar feeding rations in line with their stage of pregnancy/save veterinary fees.   2

    (ii) 18–22 days   1 This is time interval between the two peaks of lambing in synchronised ewes.   1

d) Given an inert substance instead of progesterone/no hormone given; Otherwise kept under same conditions as experimental group/ Valid example of controlled variable e.g. food supply.   2

Total marks = 15